AF589385

ENDOCRINE GLANDS OF THE FISH, *NOTOPTERUS NOTOPTERUS*

The Author

Dr. Raghavendra S. Kulkarni did his M.Sc. in Zoology from Karnatak University, Dharwad and Ph.D. in Zoology from Banaras Hindu University, Varanasi during 1980 with fellowship from UGC/JRF and CSIR/SRF/PDF under the guidance of Prof. A. G. Sathyanesan. He has teaching and research experience of 35 years. His specialization is Reproductive Biology and Comparative Endocrinology. He was selected under Emeritus Fellowship from University Grants Commission, New Delhi standing second in the order of merit at National level and He is presently working as distinguished UGC Emeritus Professor of Zoology at the Department of studies in Zoology, Gulbarga University, Kalaburgi, India. He has published more than 140 research publications in reputed journals and successfully guided 12 students for Ph.D. degree and 20 students for M.Phil degree on various aspects of fish endocrinology and reproductive biology. He has handled Research Projects from ICAR, UGC and has participated several National & International conferences in India and abroad, chaired the sessions also visited Japan, Belgium, Germany, France, Singapore, Nepal and Ethiopia for academic visits. He is recipient of Eminent Scientist (2005) and Best Scientist award (2007) from National Environmental Science Academy, New Delhi and he has received Dr.Abdul Kalam Life time Achievement National Award from Krist Foundation, Bangalore. He is recipient of fellowships with title from Zoological Society of India (FZSI), Academy of Environmental Biology (FAEB) and Society for Science and Environment (FSSE). He is life member of ZSI, AEB and Indian Science Congress Association.

ENDOCRINE GLANDS OF THE FISH, *NOTOPTERUS NOTOPTERUS*

Dr. Raghavendra S.Kulkarni
Ex Professor & Chairman
UGC Emeritus Fellow
Department of Studies in Zoology,
Gulbarga University, Kalaburgi-585106
Karnataka, INDIA

2018
Scholars World
A Division of
Astral International Pvt. Ltd.
New Delhi – 110 002

ISBN 9789387057586 (International Edition)

Publisher's Note:

Every possible effort has been made to ensure that the information contained in this book is accurate at the time of going to press, and the publisher and author cannot accept responsibility for any errors or omissions, however caused. No responsibility for loss or damage occasioned to any person acting, or refraining from action, as a result of the material in this publication can be accepted by the editor, the publisher or the author. The Publisher is not associated with any product or vendor mentioned in the book. The contents of this work are intended to further general scientific research, understanding and discussion only. Readers should consult with a specialist where appropriate.

Every effort has been made to trace the owners of copyright material used in this book, if any. The author and the publisher will be grateful for any omission brought to their notice for acknowledgement in the future editions of the book.

Published by : **Scholars World**
A Division of
Astral International Pvt. Ltd.
– ISO 9001:2015 Certified Company –
4736/23, Ansari Road, Darya Ganj
New Delhi-110 002
Ph. 011-4354 9197, 2327 8134
E-mail: info@astralint.com
Website: www.astralint.com

Preface

This manuscript entitled "Endocrine glands of the fish, *Notopterus notopterus* "has been prepared keeping in view of not availability of methodology being adopted for endocrine studies in fish particularly and this complete manuscript is a outcome of my over a 30 years period of my teaching and research experience for post graduate and research students. A comprehensive account of a particular fish species on endocrine glands was not available and I thought that this could help our young researchers carrying research work on comparative endocrinology and more specifically with reference to fish endocrinology and reproduction. With my experience in handling the fish tissue during my doctoral and post doctoral studies at Banaras Hindu University, Varanasi under a eminent fish endocrinologist Professor A.G.Sathyanaesan made me to continue my previous research work and further carryout research work on fish endocrinology and biology of reproduction here at Gulbarga University, Kalaburgi on my appointment in the Department of studies and Rresearch in Zoology. I feel and hope the methodology given in this manuscript may help and inspire to carry out the research for future group of students joining research in the field of fish research in the coming years.

I have to thank all my research students joined under my guidance to undertake research work leading to their Ph.D. degree in the department of studies and research in zoology, Gulbarga university, Kalaburgi. It was my memorable experience with them to work together on The various research topics related to fish endocrinology and biology of reproduction in the fish, Notopterus notopterus which made me to write this manuscript. I have to thank all my research students specialy Dr. B.J. Koppar, Dr. M. Patil, Dr. Mohd. Rafiq, Dr. Shankar, Dr. Shinde, Dr. Sudarshan, Dr. Sathesh, Dr. Khaleel Patel, Dr. Sahana, Dr. Vinayak , Dr. Pruthviraj, Dr. Ravikiran and my colleagues Prof. K. Vijaykumar, Chairman of the Department of Zoology, GUK and also Dr. Murali Jadesh for their help. This manuscript is written while I was working as a Emeritus Fellow selected by the University Grants Commission, New Delhi for two years during the years of 2015-17 and I am grateful to UGC for giving this opportunity to carryout work related fish research, also Gulbarga University for providing facilities to undertake research work at The Department of Zoology, GUK.

I extend my gratefulness to my teachers Prof. A.G. Sathyanesan, Prof. V.B. Nadkarni and Prof. S.K.Saidapur, for their inspiration and guidance.

Dr. Raghavendra S.Kulkarni

Contents

Preface v

Introduction to Fish Endocrinology ix
About Fresh water Fish, Notopterus notopterus

1. **Hypothalamo-hypophysial axis in the Fresh Water Fish, *Notopterus notopterus*** 1
   - *i. Structure of hypothalamus*
   - *ii. Hypothalamo-neurohypophysial neurosecretory system*
   - *iii. Hypothalamo-hypophysial Vasculature*
   - *iv. Structural organisation of the pituitary gland*
   - *v. Hormones of the pituitary gland*
2. **Structural Organisation of Pineal Gland and Melatonin** 35
3. **Structure and Distribution of Thyroid Gland** 39
   - *i. Thyroid hormones and its effects*
4. **Structural Organisation of Adrenal Components and Corpuscles of Stannius** 43
   - *i. Adrenal steroids and their effects*
   - *ii. Structural organisation of corpuscles of Stannius*
   - *iii. Hormone of the corpuscles of stannius*
5. **Structure of the Pancreas and its Hormones** 59
6. **Gonads and Gonadal Hormones** 63
   - *i. Structure of testis and testicular cycle*
   - *ii. Testicular steroid synthesising cellular sites*
   - *iii. Testicular steroid hormones*
   - *iv. Structure of the ovary and ovarian cycle*
   - *v. Ovarian steroid synthesising cellular sites*
   - *vi. Ovarian steroid hormones*

References 77

Introduction to Fish Endocrinology

About Fresh Water Fish, *Notopterus notopterus*

Introduction to Fish Endocrinology

The knowledge on fish biology particularly on morphology, morphometry, length-weight relationship, condition factor, reproduction, food and feeding habit, etc. is of outmost important not only to fill up the lacuna of our present day academic knowledge but also in the utility of the knowledge in increasing the technological efficiencies of the fishery entrepreneurs for evolving judicious pisciculture management. For developing fishery, it is necessary to understand their population dynamics, how fast they grow and reproduce, the size and age at which they spawn; their mortality rates and its causes, on what they prey upon along with other biological processes including endocrine aspects such as structural organization of endocrine glands, cellular identification, hormones from different endocrine glands and their physiology of a particular fish species.

There are many isolated disciplines in fish biology, of which the study of morphology is inseparably related to study of the mode of life of the organism. In fact, the size and shape are fundamental to the analysis of variation in living organisms and morphological variations even in the same species most often related to the varied environmental factors. Fishes are the first vertebrates with jaws. They are cold-blooded animals that breathe by means of gills, live in water and move with the help of fins. There are about 36,000 species or more, which represent the 40% of the total vertebrates present. Fishes have evolved during Ordovician period and widely distributed during Devonian period, which is known as 'Golden age of fishes'. Fishes are economically very important group of animals. They are used as food throughout the world and the fish liver is the main source of liver oil containing vitamin A and D. Body oils of fishes are externally used in soap industry and tanneries. Beautiful colored fishes are the present craze to have them in Aquariums.

Endocrine systems usually control long-term activities of target organs and also physiological processes such as digestion, metabolism, growth, development, reproduction etc. The endocrine system includes endocrine glands, which are distributed in various regions of the body of fish. Endocrine glands of increasing complexities are found in elasmobranches and osteichthyes. Elasmobranches (sharks) possess well developed endocrine glands but these show some interesting differences from those of higher chordates. However, osteichthyes have endocrine glands rather more similar to higher chordates. The difference between fish and mammalian endocrine glands is probably due to the development and modification of various body systems in these two classes, and also due to exogencies of an aquatic mode of life. Mammalian endocrine glands are well advance and well studied but fish endocrinology is limited to the work on its influence on chromatophores, action of sex cells, function of pituitary and thyroid and control on migration.

Unlike nervous system, the endocrine system is basically related to comparatively slow metabolism of carbohydrate and water by adrenal cortical tissue, nitrogen metabolism by adrenal cortical tissue and thyroid glands and the maturation of sex cells and reproductive behavior by the pituitary gland and gonadal hormones.Hormones strictly speaking stimulating substrates and act as body catalysts. They catalyze and control diverse metabolic process. The hormones are not only catalysts for all metabolic process; they are living proteins that direct the life force into over basic biochemical and metabolic process.

The hormones help transform store energy and active other physiological process including their participation in reproductive activities. The hormones as a class of compounds have three major functions in the body namely integrative, regulative and morphological. A particular hormone may be involved in only one function. The integrative action is possible because the hormone travel through blood from a source to their site of actions enabling the body to function as a unit response to stimuli. Hormones are involved in the control of the biochemistry, physiology and behavior of all species studied so for. They coordinate complex physiological processes such as development, growth, metabolism reproduction and behavior. The role of hormones in the regulation of various aspects of reproductive behavior has been reported to emphasize the two-way nature of the relationship between the endocrine system and the biological and physical environment. Not only does the endocrine system regulate the behavioral responses necessary for successful reproduction, but it is also responsive to social and other exogenous stimuli.

The gametogenic cycles of circulating sex steroid hormones in fishes and their relation to gonadal maturation and its environmental control, have most often been studied in primitive freshwater teleosts, particularly salmonids and cyprinids. The seasonal changes in water, fat and protein content of muscle and liver found in several species have been related to growth of gonads and other processes associated with spawning and all these processes are regulated by hormone. In all oviparous species including fish, the process of vitellogenesis is an important step of ovarian development in which the future embryonic development depends. A number of such studies concerning hormonal changes during vitellogenesis maturation and ovulation have been reviewed. Ovarian steroid biosynthesis has been widely studied in several species of teleost fish. The mechanism implicated in the steroidogenesis are similar to those described in the mammals; *viz.*, synthesis of pregnenolone from cholesterol or from acetate via cholesterol; synthesis of C19 steroids from pregnenolo-e via the 5-ene and / or the 4-ene pathway synthesis of the C18 steroids by aromatization of androstenedione and / or testosterone; and synthesis of 5 - and / or 5 - reduced compounds from pregnenes androstenes.

It is fact that gonadal hormones have been implicated in one way or another in reproductive drive. Special behaviors are made possible by the presence or absence of hormones at particular point. These hormones render the animals sensitive to certain stimuli from the environment. It has been assumed that reproductive behaviors must be regulated by hormones in the blood and the reproductive and some other behaviors in the fish are associated with hormonal control and regulation. Though, the brain (including pituitary), and gonads are necessarily involved, but the gonads and gonadal hormones are of much more importance, gonadal hormones have been found to affect certain non-reproductive behaviors also. Gonadal hormone show several marked effect on behaviors.

The observation on Gasterosteus concluded that a complete behavior related to reproductive cycle can be expressed whereas, combination effects of gonadotropic and gonadal hormones are taken into consideration. The sexual behavior in guppy is under direct control of the pituitary and the gonads exert regulatory influence. It has been described an indication of influence of FSH and LH in cichlids, on aggressive behavior along with the effects of prolactin. The sex steroid levels in gonads and peripheral blood plasma useful indicators of steroidogenic secretion during particular stage of the sexual cycle and in association with changes in gonad condition, have fostered in understanding of the endocrine control of reproduction in teleosts.

Determination of steroid hormones and gonad condition during the annual reproductive cycle of various teleosts has been reported. The hormone estradiol-17β was first identified in ovaries of the trout, *Salmo irdeus* and the carp, *Cyprinus carpio.* Since then this hormone along with estrogen has been identified in the gonads of several species of teleosts. Estradiol-17β is the major ovarian steroid of most teleosts and are involved in the synthesis of vitellogenin in the liver. This protein (Vg) synthesized in the liver is transported in the blood to the oocyte where it is absorbed and deposited as yolk. In most of the species studies so far, estradiol-17β is the primary regulator of hepatic yolk synthesis. It is a precursor molecule for the primary egg yolk proteins of all egg-laying animals. In addition to this, estrogens regulate carbohydrate and lipid metabolism. The growth and development of vitellogenic eggs is also closely accompanied by the presence of high level of 17β-estradiol in the plasma of teleosts. Estrogen treatment to fish also appears to result in a general gearing up of metabolism to provide the large amount of energy and reducing power (NADPH) necessary for protein and lipid synthesis. In the oocytes of *Xenopus laevis* it has been reported that 99% of the yolk protein is derived of hepatic origin yolk protein (vitellogenin). The levels of the hormones in seasonally breeding animals have been extensively worked out. In fishes hormone also exhibit similar change with respect to season and the reproductive stage such studies showing hormonal changes during different stages of reproductive cycle have been reported in the fish. Seasonal variation of the steroid hormones testosterone (T), 11-keto-testosterone (11KT), 17β-estradiol (E2) and cortisol have been studied in the plainfin midshipman fish, *Porichthys notatus* indicating that the sex specific peaks of steroid hormones levels in male and female may have some differential fluctuations related to the physiology, reproductive behaviour and vocal communication. Such changing levels of hormones have been implicated for storage and utilization of biochemical compositions of tissues specially muscle and liver tissues since they are site of action of various hormones produced in fish.

The detailed study has been undertaken and presented in this manuscript on the fresh water fish, *Notopterus notopterus* a locally available fish on different endocrine glands, including their structural details, location, hormones of the endocrine glands and specific hormonal effects.

About Fresh Water Fish, *Notopterus notopterus*

The fresh water teleost *N. notopterus* is selected for the present study. This fish is locally available in ponds, tanks and rivers in and around Gulbarga city. The fishes were collected from Bheema river situated at Katti Sangavi, which is about 30 Km away from Gulbarga. This fish is popularly called as "Chambari". The weight of the fish ranges from 90 grams to 150 grams and length ranges from 18 cms to 28 cms.The colour of the fish on the dorsal side either coppery brown with grayish and slivery colour on the ventral side. Body is highly compressed laterally, head is small with large mouth. A dorsal fin, pectoral and pelvic fins are small in size, anal fin is much elongated and confluent with caudal fin, which is very small or aborted (Gephyrocercal).

The fish feeds on insects, small fishes, crustaceans and some young roots of aquatic plants. The fish do not exhibit sexual dimorphism externally and is active during twilight and night. Although it is not a commercial important fish. However, it is gaining importance as food fish in the local area.Breeding takes place in stagnant and running waters during rainy season. Eggs are laid in small clumps on submerged vegetation.

Classification : According to Talwar and Ihingran (1991), the fresh water fish *Notopterus notopterus* is classified as follows:

Phylum	Chordate
Subphylum	Vertebrate
Super Class	Gnathostomata
Grade	Pisces
Class	Osteichthyes
Subclass	Actinopterygii
Order	Osteoglassiformis
Sub order	Notopteroidei
Family	Notopteroidae
Genus	*Notopterus*
Species	*notopterus*

Showing the photograph of the fish, *Notopterus notopterus.*

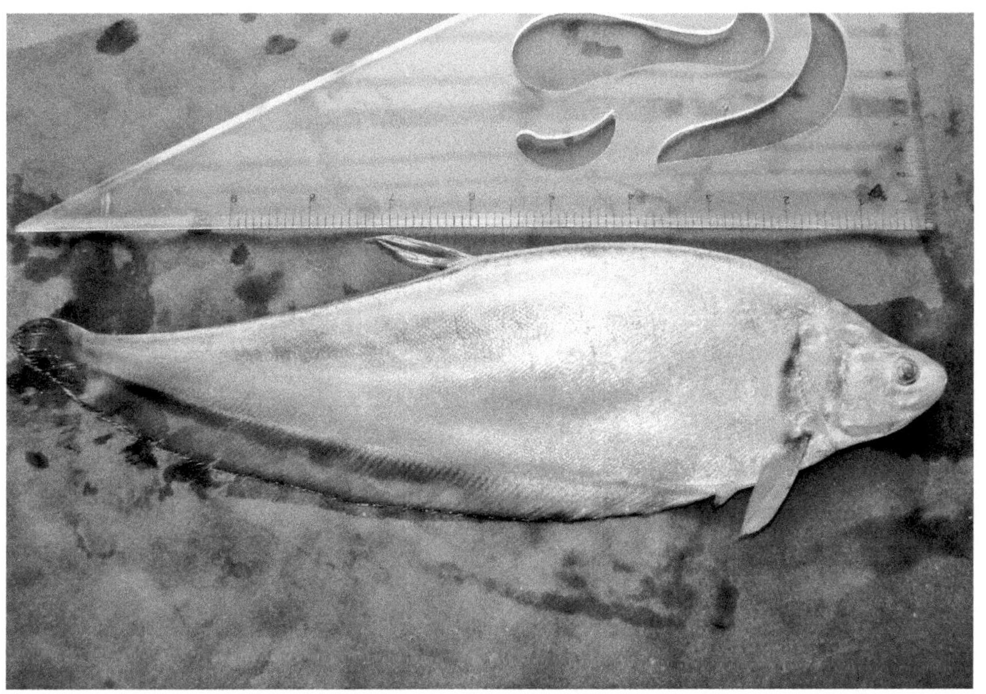

Chapter - 1

Hypothalamo-hypophysial axis in the Fresh Water Fish, *Notopterus notopterus*

Structure of Hypothalamus

The hypothalamus of the fish, *Notopterus notopterus* consists of III Ventricle with infundibular recess, nucleus preopticus (NPO) on either side of the III rd ventricle at the anterior side of the hypothalamus, the neurosecretary tracts on either side joining at the stalk of the pituitary, the nucleus lateralis tuberis (NLT) and ramified capillaries at the median eminence forming neuro-adenointerface vasculature. Some of the neurosecretary neurons terminating at the median eminence. A saccus vasculosus is present below the pituitary gland in the fish, *Notopterus notopterus.*

Hypothalamo-neurohypophysial Neurosecretory System

The hypothalamo-hypophysial neurosecretory system and vasculature studied in the freshwater fish *Notopterus notopterus.*

The hypothalamo-neurohypophysial (HN) complex of *N. notopterus* consists of nucleus preopticus (NPO) its axonal tracts and neurohypophysis (NH). The nucleus preopticus (NPO) is paired situated on either side of third ventricle in close proximity to the ependyml layer. In the ventral view of in situ bulk stained brain, the entire HN complex can be demonstrated. In such a bulk preparation the NPO is seen in a group as a rounded structure. However, in a cross sectional view, the intact NPO is vertically oblong and in para sagittal view it looks dorso-ventrally elongated, thus the whole NPO differs in its configuration when viewed from different planes.

Although the NPO is vertically set in the para sagittal view, microtome sections show scattered neurons extending backwards from the dorsal extremity of the NPO. These scattered cells form almost a straight line which joins the bulks of the NPO cells roughly. Thus the NPO in this species is in the form of 'arch' shaped in the median sagittal section. The dorsal half of the NPO which is formed of larger neurons is called as pars magnocellularis (PMC) and the ventral half and dorsally localised consisting of smaller neurons is termed as par parvocellularis (PPC). The NPO is situated antero dorsal to the optic nerve, it is situated antero dorsal to the optic nerve, it is clearly seen in microtome para sagittal section. The majority of the NPO cells are oval or oblong in shape having round nuclei. Bipolar and multipolar neurons are common in parvocellularis (PPC). The neurons of both PMC and PPC contribute to the fine beaded axons for the formation of left and right tracts and these two tracts unite to form a common tract which enters the pituitary (Diagram). In the sagittal view of the bulk preparation of the brain. the neurosecretory tract is seem to be coming form the PPC region because both PMC ad PPC axons contribute to the formation of the main neuro secretory tract. This neuro secretory tract enters into the pituitary in between the RPD and anterior infundibular base (Diagram).

This neuro secretory tract enters into the pars intermedia (PI) and remifies. The ramified neuro secretory tract have pervi vascular endings smaller neuro secretory axons are seen among the cells of the PI due to greater accumulation of neuro secretory material the entire PI is stained dark in bulk preparation and in the tissue sections the neuro secretory tract entering into PI is clearly seen. The nucleus lateralis tuberis (NLT) cells extends from the base of the infundibular stack to the midway between the pituitary and optic chiasma along the floor of the brain. The nucleus lateralis tuberis NLT cells are more in adjacent to the infundibular base and they are gradually decreased in their number towards optic chiasma.

Diagrammatic representation of hypothalamo-hypophysial neurosecretory system in fish:

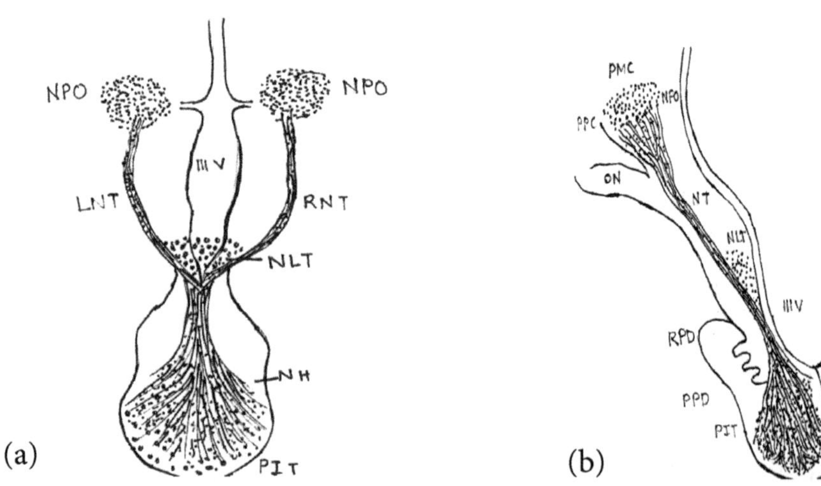

N. notopterus. (Ventral View).

N. notopterus. (Saggital View)

Fig.1.1: (a) and (b)

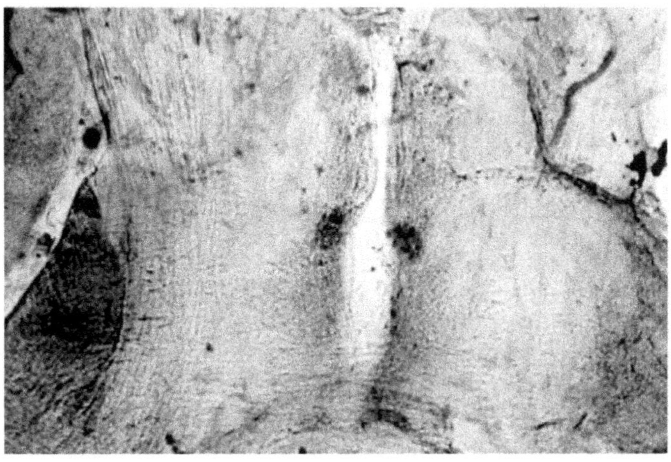

Fig.1.2: Section of ventral view of the hypothalamo-hypophysial complex. NPO stained deeply with AF. Bulk staining technique.

Showing section of ventral view of the hypothalamo- hypophysial complex. NPO and PIT stained deeply with AF-Bulk staining technique.

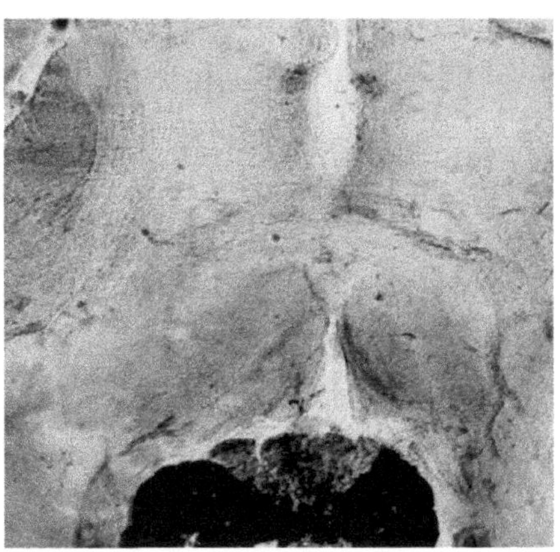

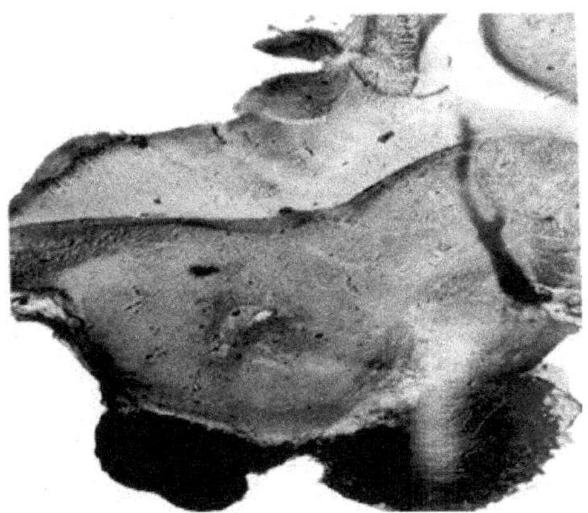

Fig.1.3: A thick sagittal section of the brain with intact pituitary and saccus vasculosus showing hypothalamo–hypophysial complex AF positive hypothalamo- neurohypophysial tract. Bulk staining technique.

Hypothalamo-hypophysial vasculature

With India ink perfusion technique the blood vessels entering into the hypophysis has been studied. The artery entering in the region of anterior to the hypophysis divides into two branches One branch (hypothalamic artery) gives its branches forming capillaries in the region at the anterior ventral part of the hypophysis (median eminence-ME). The other blood vessel, hypophysial artery enters the pituitary and largely contributes to the interface vasculature lying in between the neurohypophysis (NH) and pars distalis. The blood vessels from the interface network irrigate pars distalis which is very clearly seen. In the frontal sections and in sagittal sections the vasculature forming a neuro-adeno interface vasculature comprising of network in between neurohyposial (NH) and pars distalis are seen. The portion of the hypothalamus anterior to pituitary where nucleus lateralis tuberis (NLT) cells are present show capillary formation and some of these capillaries directly enter the pituitary where they ramify in the glandular pituitary.

The neurosecretory AF positive axons entering into the pituitary are clearly visible in these regions where capillary network formation is present in between neurophpophysis and pars distalis. Superimposing on the blood vessel may come in contact with neurosecretory tract. It is possible that neurohormones might get access to the circulation at this region. In this fish neuro-adeno interface vasculature is very clearly visible in addition to capillaries in the region of anterior hypophysis. The venous blood collected by complicated network of blood vessels finally joins posteriorly into the hypophysial vein. The vascular connection existing between the pituitary and saccus vasucloses needs further study. In frontal and sagittal section of India ink injected preparation showed that the nucleus preopticus (NPO) region is highly vascularized.

The hypothalamo-hypophysical vascularization in *Notopterus notopterus* resembles with that of *Channa punctatus* consisting of arteries contributing to the vascular network of hypothalamus in the region immediately anterodorsal to rostral pars distalis (RPD) which is continued into the neurohypophysis (NH). This network was reported to be comparable to the primary capillary plexus of higher vertebrates. However, they do not form prominent short and long loops as present in mammals but their meshes run into one another with a few portal vessels entering the RPD. The clear capillary plexes and their irrigation in the region of anterior NH forming neuro-adenointerface vasculature.This region has some AF positive neuronal endings and also has neurosecretory tract, suggesting the secretory products might get into the vessels and reach the adenohypophysial cells. This study forms a morphological evidence to demonstrate the presence of hypothalamo-hypophysial portal system in this fish. The functional aspect to demonstrate the presence of portal system seen outside the pituitary of *N. notopterus* like that of terapods needs further study. However, the neuro-adeno interface vasculature is prominently seen in *N. notopterus*. Such a interface vasculature has been reported in variety of other Indian fishes. *Clarias batrachus, Heteropnustes fossilis, Channa punctatus, Bagarius bagarius, Rita rita. Trichogaster fasciatus, Labeo rohita, Puntius sophore. Cirrhinus mrigala* and *Mystus vittatus.* In addition to neuro-adeno interface vasculature tetrapodan type of pituitary portal system has been also demonstrated in these fishes. Recently reported a terapod like hypothalamo-hypophysial portal system in the teleost. *Megalops cyprinoides,* in which the complicated loops of the primary capillary plexus are seen on the perikarya of the NLT and the preoptic hypothalomo-neurohypophysial neurosecretory tract. As early as in 1969 Ball and Baker, have reviewed the blood supply to the teleostean pituitary gland and reported that the scattered observations might be interpreted as indicating some kind of portal connection between hypothalamus and adenohypophysis, the important point to emphasize is that in no teleost has any close association between hypothalamic neurosecretory nerve endings and blood capillaries been observed outside the pituitary itself. Thus whether or not these various hypothalamic capillary contributions to pituitary vascularisation represent as it were a foreshadowing (or reminiscence) of the portal condition, there is no reason to regard them as functional teleostean portal systems". Blood from the secondary plexus drains posteriorly into venous system which colasces into one large vein existing in the pituitary. Further, he explains in some species in which vestigial hypothalamo-hypophysial portal system exists. The teleosts forms a major group amongst the Antenopterigians, the other infraciasses of Actinopterigian group are Chondrostei (e.g., *Polypterus* and *Acipenser*) and Holostei (e.g., *Amia* and *Lepisosteus*), which exhibit a functional blood portal system and does serve to transport neurohormones to the pars system and does serve to transport neurohormones to the par distalis. Perks (1969) in his review observed "the median eminence (ME) and pituitary portal system are common features of the neurohypophysial system from the level of the primitive hagfish up through the vertebrates and the teleosts have developed direct neural contacts with the adenohypophysis and it is possible that the portal system became absolute and was lost". With the available literature on the presence of hypothalamo-hupophysial portal system in some teleost reported by Sathyanesan and his group. The present study on the vasculature of *N. notopterus* showing capillary plexus at the region anterior to pituitary in the hypothalamus and a prominent neuro-adeno interface vasculature observed is an additional support to demonstrate morphological evidence of presence of a clear vascular link between hypothalamus and hypophysis in *N. notopterus.*

To demonstrate the functional aspect of hypothalamo-hypophysical vascular (portal) system in *N. notopterus* needs further analysis and detailed study.

The hypothalamo-neurohypophysial (HN) complex of teleosts comprises of nucleus preopticus (NPO), their axonal tracts, nucleus lateralis tuberis (NLT) and neurohypophysis (NH). Although the teleostean HN complex has been extensively studied and reviewed by several investigators. The structural details of HN complex has been studied in a few species using *in situ* staining techniques. In some Indian teleostean fishes, Sathyanesan and his group have studied the structure of the hypopthalamo-neurohypophysial system using *in situ* staining procedures, describing that the distributional pattern of HN complex remains same, amongst the Indian cat fishes studied for HN complex are, *Clarias batrachus* , *Heteropneustes fossilis, Rita rita* , *Ailia coila, Ompak bimaculaus* and *Mystus vittatus.*

Different species are known to have NPO of different shapes. They may be fusiform, rounded, polygonal and flask shaped. In his extensive review, Perks (1969) has described the course of the neurosecretory tracts which pass out of the nucleus in an irregular diffuse manner, often leaving in a lateral direction and then bending ventrally before forming a discrete tract in the infundibular floor. In *Nangra punctata* , the intact NPO exhibit different shapes when viewed from different planes. Both Pars parvo cellularis (PPC) and pars magnocellularis (PMC) contribute to the formation of right and left main neurosecretory tract which do not enter the pituitary directly, on the other hand they give rise to several pairs of lateral tracts, they join to form a pair of median tract which enters the pituitary, *in situ* studies in *Channa punctatus,* more number of neurosecretory tracts were seen independently running along the lateral aspect and enter the NH. In *Micrognathus aculeatus* also these tracts runs laterally but they are loosely consolidated in left and right tracts. In *C. batrachus* the NPO cell appearance varies when viewed from different planes. In ventral view they appear like a paired transverse bands situated on either side of the third ventricle and in cross sectional view the whole NPO is in the form of U shape, in side view, looks like a vertical band. In *Rita rita* The NPO is in the form of two rounded bodies located on either side of the third ventricle and in thick cross sectional view the entire NPO shows that they are in the form of two vertically set elongated bodies. In this species main neurosecretory tracts are composed of loosely set axons for about one third distance towards the hypophysis later they become more compact and give off five or more pairs of lateral tracts which are less prominent in *C. batrachus.* The structure of the HN system of *M. vittatus* is largely comparable to that of other cat fishes described earlier by Sathyanesan and co-workers. In *N. notopterus* NPO cells are oval and oblong shaped when viewed from different planes. Since the shapes of NPO cells are varied and whole NPO extends into the hypothalamus to a varying depths, all its components cannot be seen in one plane, the NPO exhibits like a arch shaped bend in sagittal view.When viewed from the ventral side, the NPO is in the form of rounded structure as in *Rita rita* and it is situated on either side of the third ventricle. The shape of the whole NPO can be also comparable to that of *Nangra punctata.* The NPO cells are having close proximity to ependymal layer as in some cat fishes described.

The third ventricle and ependymal cells of *N. notopterus* has a wide and direct contact with the neurohypophysis. This morphological contact of the ependymal cells with the neurohypophysis and also with the neurosecretory tract has been reported to have some physiological significance.

Cerebrospinal fluid and ventricular system have been known as the path way of neuro endocrine integration. There is a general agreement that AF positive peptidergic fibres originate from the NPO and the negative ones from NLT. It has been suggested that atleast five morphologically distinct fibres innervate adenphypophysis in some fishes. The nerve tracts entering into neurohypophysis of *N. notopterus* gave positive reaction to AF which is originating from NPO. The nucleus lateralis tuberis (NLT) cells are AF negative and extend from the base of the anterior portion of the infundibular base close to the pituitary, midway between optic chaisma and pituitary. The histochemical study of these cells is needed for further analysis. In *N. notopterus* NPO, their axonal tracts entering into the PI along with NLT forms the HN complex. This HN complex of *N. notopterus* is more or less similar to that of *Channa punctatus.*

Showing sagittal sections of the brain with inta pituitary and saccuss vasculosus after perfusion of India ink and enlarged showing the vasularisation in the pituitary with ramification of blood vessels.

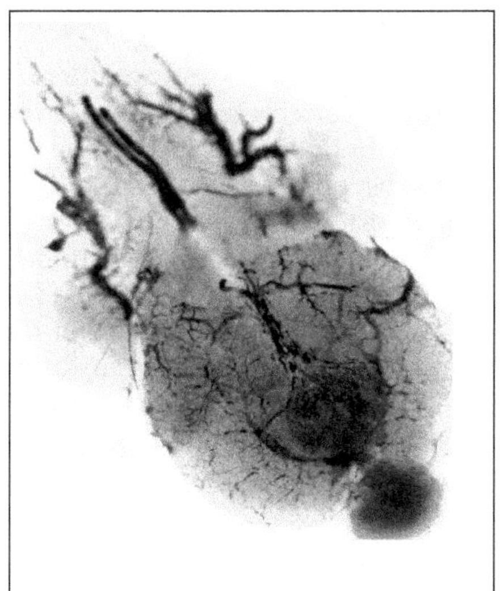

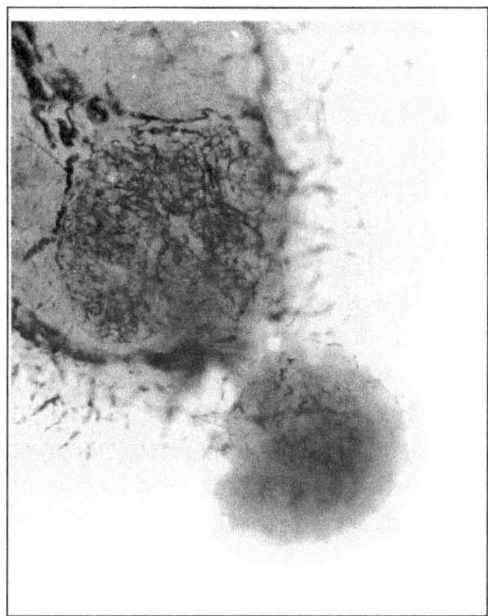

Fig.1.4: Frontal view of the brain with intact pituitary after vascular perfusion of India ink showing blood vessels in the ME and pituitary.

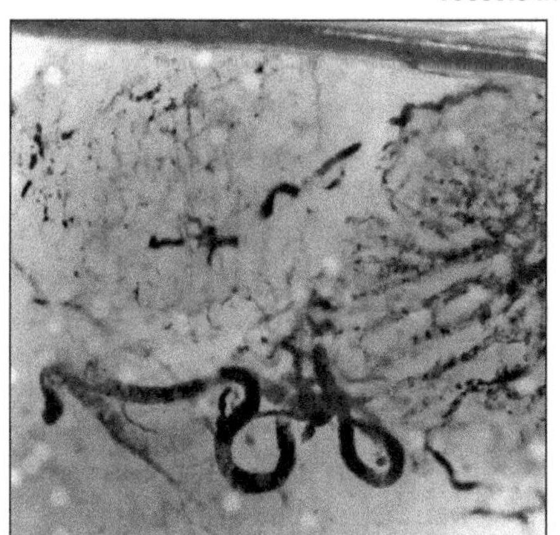

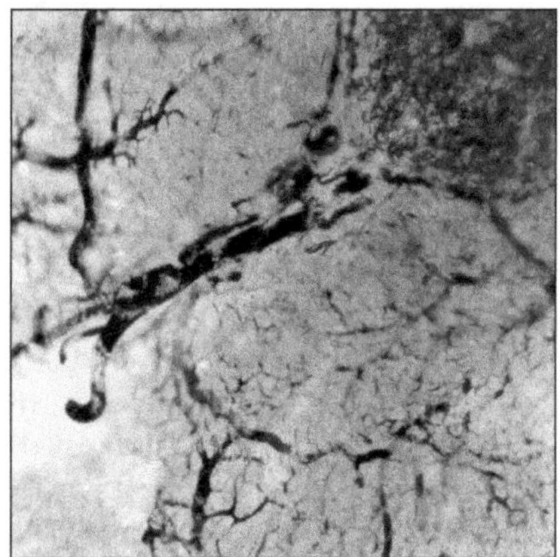

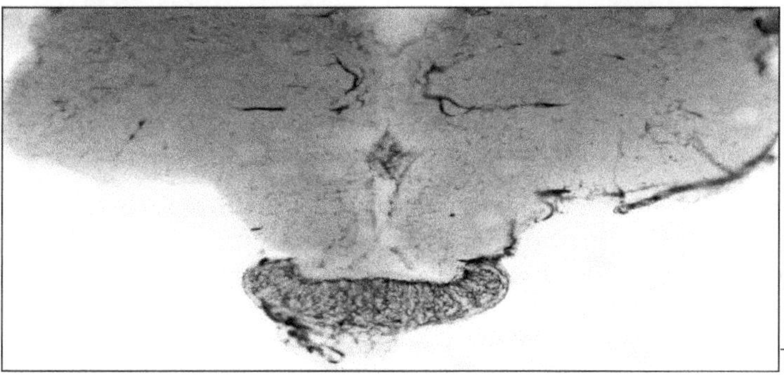

Fig.1.5: India ink injected frontal section of the brain with intact pituitary showing neuro adeno interface vasculature in the fish, *Notopterus notopterus.*

Structural Organisation of the Pituitary Gland

The study on the distributional pattern of cell types in the teleostean pituitary gland has attracted greater attention than any other vertebrate, because of the ability of the cells to respond to various staining methods and also experimental responses.Though morphological structure of the teleost pituitary has been studied, the details of the fine structure still remain open for investigation especially amongst Indian fishes of which only a few species have been studied.The teleostean pituitary gland is known for its remarkable structural diversity and cell types described in the adenohypophysis are shown to be physiologically equivalent to those of the mammalian pituitary gland. Due to the unique organisation and the specialization make the teleost pituitary an ideal model for the study of the relationship between nervous and the endocrine components of the neuroendocrine system , therefore, anatomical data obtained in the teleosts may contribute to the better understanding of the hypothalamohypophysial system in higher vertebrates.

The pituitary of *N. notopterus* lies on mid ventral to the brain behind the optic chiasma and it is attached to the floor of the infundibulum. The gland is without a definite stalk having a wide infundibular cavity and closely applied to the brain. The pituitary is differentiated into rostral pars distalis (RPD), proximal pars distalis (PPD) and par intermedia (PI).The RPD, PPD and PI are arranged one after the other in antero-posterior axis. The neuropypophysis is branched and possess the distinctive features with ramification in the pars intermedia (PI).The RPD is made up of erythrosine positive acidophils and PbH positive cells. The acidophils cells are also positively stained to orange-G. These acidophil cells resemble the prolactin secreting cells. The cyanophils are restricted to the dorsal region of PPD and also scattered in the form groups among the acidophils.These cyanophils are AF positive, PAS positive and analine blue positive. Pituitaries of fish having maturing and matured gonads exhibited a progressive increase in the number of cyanphils which are more concentrated dorsal aspect of the proximal pars distalis (PPD), these cyanophils are less during resting phase (October).These suggest that those cyanophils are gonadotrophs the remaining cyanophils are considered as thyrotrophs.The pars intermedia (PI) is recognised usually by its close relationship with the pars nervosa and it is the most posterior region of the gland. The main bulk of deeply stained AF positive nerve fibres of the neurohypophysis (NH) penetrate the pars intermedia (PI). The PbH positive cells are arranged around the processes of the neurohypophysis. These cells are considered as MSH secreting cells. PI is also showing small number of PAS positive cells. In all the regions of the pituitary some unstained cells are seen which are considered as chromophobes.

Seven types of cells were clearly identified, as acidophils consisting of prolactin (lactotrophs) and growth hormone secreting somatotrphs cells, basophils as gonadotrophs and thyrotrophs, lead haematoxylin (pbH) +ve cells in the RPD and PI. Some cell PAS +ve cells in the PI alongwith chromophobes distributed unevenly in the PPD. The RPD and PPD have mixed cell types showing prolactin cells, somatotrophs, gonadotrophs, thyrotrophs and corticotrophs. The prolactin cells amonst the acidophils are largely seen in the region of RPD. The PbH+ve cells seen bordering the NH in both RPD and some PPD are considered to be corticotrophs.

The cyanophils comprising gonadotrophs and thyrotrophs are more in the region of PPD.

The pituitary of fish having matured gonad during breeding period exhibits large number of intensely stained PAS and AF +ve cells in the median region of the PPD are considered as gonadotrophs.The thyrotrophs seen in the region of basal RPD are well identified by their staining affinity and their number in the nonbreeding period. The growth hormone secreting somatotrophs are seen in both RPD and PPD. The PI of *Notopterus notopterus* exhibits two types of cells as PAS and PbH + ve cells. The PAS +ve cells outnumber the PbH +ve cells. Interspersed among the GTH cells in the rostral follicles of *N.notopterus* are few small cells lying adjacent to fine channels which connect the follicular lumen with the intervascular space between the follicles. Because of their position they have been called "neck cells", and the suggestion was made by altering their size and shape, they could regulate the passage of substances along the channels as cited in the reviews. These cells in all probabilities belong to the stellate system, which are general features of adenohypophysis.These cells occur in all regions of the adenohypophysis of teleosts, their fine process penetrating between the endocrine cells as revealed in the literature. The other cell types which are identified based on the size of the secretary granules are somatotrophs and thyrotrophs, which closely associated with gonadotrophs.

Histochemical and Electron Microscopic Studies on the Pituitary Gonadotropic Cells of the Fish *Notopterus notopterus*

Gonadotropic cells are distributed in the proximal pars distalis of the pituitary in *Notopterus notopterus.* Histochemical and electronmicroscopic techniques are applied for the present study. The PPD of pituitary gland in zero age group fish has very little number of gonadotrophs. The cells are undifferentiated with indistinct boundaries. They markedly increased in number in one year aged fish, since the gonads of this age group fish shows advanced stages of growth. The gonodotrophs that largely occupy the medial aspect of the PPD increased numerically as the gonads mature and become major cell type before spawning in two age group fish. The electronmicroscopic studies reveals that, they appear oval or elongated containing a few large granules and many small lucent secretory granules and based on this two GTH cell type could be identified in the PPD containing different types of secretory granules.

The gonadotrophs of teleosts are distributed in the proximal pars distalis. However the gonadotrophs are spread to RPD at sexual maturity in various teleosts including eels and trouts and gonadotrophs increase in number and size when the fish is having the matured or maturing gonads. The histochemical and immunocytochemical studies were made on pituitary gonadotrophs to understand their nature, structure and distribution. In some species two types of gonadotrophs were reported and the work has been also carried out on the isolation characterization and purification of two types of gonadotrophs as reported in other fishes.

Although the functional cytology of the pituitary gland in teleosts has been carried out, the studies on the cell type and their functioning in relation to different age groups of fishes are not much accounted in literature. Since the gonadal maturity depends on the pituitary gonadotrophins secreted from the gonadotrophs, their related changes in number, and their activity on the age at first maturity and subsequent studies in the ageing fish will furnish a clear understanding of their role in the process of maturation of gonads in the ageing process. Hence, in the present study on the cytology of the pituitary gland with special reference to gonadotrophs is being undertaken in different age groups including their activity during different maturity stages of gonads, which will give clear information about their role in the maturation process of gonads. The ultra structural details of the gonadotrophs in aged fish (two-year) has also been made on the pituitary of *N notopterus* to identify, if any differentiation on the two types of gonadotrophs reported in other fishes can be identified. In view of this the present study on the histological and electron microscopic nature and distribution of gonadotrophs in relation to different age groups of locally available fresh water fish *N. notopterus* has been undertaken. The study would help to understand the role of gonadotrophs in gonadal maturation so as to improve the efficiency of breeding on this locally available food fish. Since fecundity of a fish is directly related to potency of the pituitary glands of both donor as well as recipient.

Present investigation is aimed to study the gonadotrophic cells using specific staining techniques. Hence, after exposing the roof and sides of the brain, the heads were fixed in Bouin's fluid, as it is not possible to remove the pituitary intact with brain, they were decalcified by keeping them for longer time in fixative (15 days). After removing the brain with intact pituitary from cranium, the brains were kept in fresh Bouin's fluid for 24 hrs and then dehydrated by passing through 70%, 90 % and 100 % alcohol. Microtome sections (five microns thick) of the brain with pituitary were taken in parasagittal, frontal and transverse planes and stained. The following staining methods were applied for the study of gonadotrophic cells.1.Aldehyde fuchsin (AF) preceded by acid permanganate oxidation using fast green – chromotrope 2R as the counter stain. 2. Lead haematoxylin (PbH) , preceded by PAS. 3.Periodic acid (PAS) orange G (OG) procedure. 4.Cleveland and Wolf's trichome method. 5.Aldehyde thionine – PAS – Orange G method.

Electronmicroscopy

Fixation: The pituitaries were fixed in glutaraldehyde (because its capacity to stabilize most of the protein without coagulation) following fixation in the primary fixative, the tissues are washed in 0.1 M buffer. The tissues were post fixed in 1% OSO_4 for 1-2 hours at 4^0C. The tissue is embedded in embedding medium (Epon 812/ Araldite Cy212).To obtain a high resolution, the thickness of the specimen should be about 600 A (60 nm). Such thin sections are obtained on ultra microtome. The tissue is double stained with uranyl acetate followed by lead citrate. The ultra thin sections placing in copper grids observed under electron microscope at NIMHANS, Bangalore.

The gonadotrophs or GTH cells in the pituitary gland of *N. notopterus* have a wide distribution in the region of proximal pars distalis (PPD). These gonadotrophs were differentiated from the thyrotrophs (as they stain alike) by comparing the pituitary gland stained during pre-spawning and post spawning period.

The gonadotrophs are abundantly distributed in the medial region of PPD as these basophils stains intensely to AF during pre-spawning period. Hence are identified as gonadotrophs.The pituitary gland of zero age group fish has few numbers of gonadotrophs and they are undifferentiated with indistinct boundaries. The gonadotrophs markedly increase in the (PPD) region of pituitary in one year age group fish and also show advanced stages of growth. Since gonadotrophs are involved in the process of growth and maturation of gonads, the elder ones (1 and 2 aged) need more GTH as compared to younger ones (0 age and 0^+ age). Hence the gonadotrophs are abundantly distributed in the PPD of pituitary in fish having two year age. The gonadotrophs are intensely stained in the prespawning period. These cells undergo degranulation with weak staining after spawning in the same age group fish. In one year aged post spawned fish the gonadotrophs get degranulated indicating the hormone release during spawning activity. Gonadotrophs in two year aged fish stains positively with PAS, AB, AF, ATH and aniline blue. They are distinguished with thyrotrophs as they are positive to PAS, AF and aniline blue, but thyrotrophs have less affinity for AF, AB than gonadotrophs and a stronger affinity for PAS. The gonadotrophs are more in number and in size and their intense staining, since the fish is having maturing gonads.

The ultrastructural characteristic of the GTH cells reveals that they are oval or elongated containing few large granules and many small lucent secretory granules. Large bodies are PAS positive, AF positive and aniline blue positive and presumably are lytic granules. The cells have well developed RER and Golgi complex the nucleus is crescent shaped with prominent nucleolus. The nucleus is situated centrally or eccentrically depending upon the position of the cell. The GTH cells have complete or partial loss of glycoprotein secretory granules and development of large vacuoles in conjunction with them, indicating synthetic and secretory activity. The pituitary gland of *N. notopterus* was processed during August in which the fish had ripe spawning gonad, the ultrastructural details of gonadotrophs during this period shows characteristic of two different cellular inclusions. One type of gonadotroph containing numerous dilated cisternae of the ER (Endoplasmic reticulum) and small type of secetory granules. The diameter may be around 100-300 nm, whereas another type of gonadotrope although contain similar dilated cisternae of the ER and small type of secretory granulas, they also contain large globules of diameter around 500-4000 nm. The first type of cell described above may be GTH-I and the second may be GTH-II gonadotrope. However, there is overlapping of each other cells due to changes in number and size of the granules, globules and cisternae of the granules endoplasmic reticulum (GER). A clear distinction to make on the existence of two types of gonadotrophs in this species (*N. notopterus*) needs a study on the immunocytochemistry using EM. Interspread among the GTH cells in the rostral follicles of *N. notopterus* are few small cells lying adjacent to fine channels which connect the follicular lumen with the intervascular space between the follicles. Because of their position they have been called "neck cells", and the suggestion was made by altering their size and shape they could regulate the passage of substances along the channels as cited in the reviews. These cells in all probabilities belong to the stellate system, which are general features of the adenohypophysis. These cells occur in all regions of the adenohypohysis of teleosts, their fine processes penetrating between the endocrine cells as revealed by literature. The other cell types, which are identified, based on the size of the secretory granules are somatotrophs and thyrotrophs, which are closely associated with gonadotrophs.

The cyclic changes in secretory activity of the GTH cells during annual reproductive cycle is the most conspicuous event in the pituitary gland and are taken as a basis for their identification. The GTH cells were differentiated from thyrotrophs by comparing their distribution and intensity of staining during pre-spawning and post spawning periods of gonadal cycles. These two types of basophils stained alike with PAS, AF and aniline blue. In the post spawning phase fish pituitary only some of the cells, which are present at the basal region of PPD exhibit positive staining response while those present in the medial PPD are weak to staining. Thus the basophils of the basal PPD are considered as thyrotrophs, and the basophils present in the medial PPD are gonadotrophs. Such studies on the pituitary have been reported and those cyanophils of the middle glandular region are gonadotrophs. The distribution pattern, staining response and cytological stages of basophils during different phases of reproductive cycle of matured fish is available in the literature. The basophils distributed in the region of PPD are gonadotrophs, which are abundantly seen during pre-spawning phase of the reproductive cycle of *N. notopterus.* The pituitary of zero age group although have the gonadotrophs stainable, are very few in number and undifferentiated. These cells increased in number in one-year age group fish. They are found to be intensively positive to staining reaction and are numerically high in two-year age group fish. Thus indicating the increase in GTH cells along with their staining intensity suggest their active involvement in the development of gonads of a matured fish (in agreement with the reviews). In eel and pacific salmon two types of gonadotrophs are reported based on their colour differences and granular, size, distribution, position and vascularisation and in several other teleosts only one type of gonadotrophs are reported.

The ultrastructure, of pituitary gonadotrophs has two different types, identified based on the presence of secretory granules sizes, they seems to be similar in *N. notopterus.* Gonadotrophic (GTH) cells were very heterogeneous with regard to their size and shape. These are oval at the medial region of PPD and elongated near the basement membrane. The GTH cells found with a few round large electron dense (Lucent and Solid) secretory granules in agreement with studies of Garcia-Ayala *etal;* (1998) on the mediterranean yellowtial *Seriola-dumerilii* these large bodies are PAS positive, AF positive and aniline blue positive and presumably are lytic 'R' granules. They occur in many teleosts.The cells have well developed RER and Golgi complex (in agreement with reviews). The ultrastructural characteristics of GTH cells have been described for some catfishes including Indian catfish *Clarias batrachus.* According to these authors the most important characteristic feature of these cells is the presence of heterogenous secretory inclusions, they are either round or elongated, round globules with irregular masses identified based on their size and shape. The GTH cells of *C. batrachus* have the rough endoplasmic reticulum (RER) consisting of small dilated cisternae among the secretory inclusions and mitochondria. The Golgi complex is generally inconspicuous and is found near the nucleus. The GTH cells of *N. notopterus* also exhibit similar organisation under ultrastructural studies having heterogenous secretory inclusions rough endoplasmic reticulum (RER) with small dilated cisternae, mitochondria and Golgi complex usually appeared near the nucleus. The differentiation of two types of gonadotrophs could not be made in the histochemical studies. However, ultrastructural studies using electromicroscope indicates that the pituitary GTH cells of *N. notopterus* exhibit two different cell type are having smaller secretory granules and whereas the other having similar smaller secretory granules in addition to large globules.

Such distinction of gonadotrophs has been made in the pituitary gland of rainbow trout *Oncorhynchus mykiss* by Naito *et al.,* (1993). He describes that the salmon pituitary produces two chemically distinct gonadotrophins. The ultrastructural characteristic of GTH-I and GTH-II is studied through immuno cytochemistry using antisera against salmon GTH-I beta and GTH-II beta subunits. Those cells have dilated cisternae of the granular ER and a small number of I-beta positive granules (diameter 100-300 nm) whereas GTH-II beta immuno reactivity was found as granules (diameter 200-400 nm) and large globules (diameter 500-400 nm) in apparently differently cells (GTH-II cells). Distinct cellular distribution of GTH-I and II were maintained during gametogenesis. Although morphological characteristics of GTH-I and II cells overlapped each other due to changes in number and size of the granules, globules and cisternae of the ER, interestingly the globules in the GTH-I beta, although in the GTH-II cells, they were always stained with GTH-II beta anti serum and confirming their results reporting that GTH-I and GTH-II beta are synthesised in distinctly different cell types in the salmnoid pituitary. Recently, Matan Golan *et al.* (2016) have reported on the anatomical and functional gonadotrope networks in two teleost fish species, tilapia and zebra fish that two types of gonadotrophs based on the immunogold-lebeling and EM studies indicated that LH-producing cells were found to reside in tight clusters in peripheral parts of thePPD, whereas FSH producing cells were located more dorsall, close to the dorsal projections of the pars nervosa. In comparison to LH-cells, FSH cells were distributed more loosely throughout PPD. Similar type of differentiation and distribution is possible in The fish, *N. notopterus,* the existence of two types of cells secreting LH and FSH still needs to be confirmed using immune-histochemistry along with EM studies.

The somatotrophs have membrane bound secretary granules with clear no indication of RER or Golgi complex, the cells have few mitochondria.the granules varies in diameter observed based on the morphology and The granules are well developed. The Golgi complex showed stages in the elongation of secretary granules. The other type of cell which are associated with gonadotrophs are thyrotrophs, these cells are present in the ventral region and are intensely positive to AF and PAS , distributed in the ventral region of the RPD. The ultrastructural characteristic of the TSH cells are having dilated cisternae and varying sizes of the weekly stained smaller secretary granules.

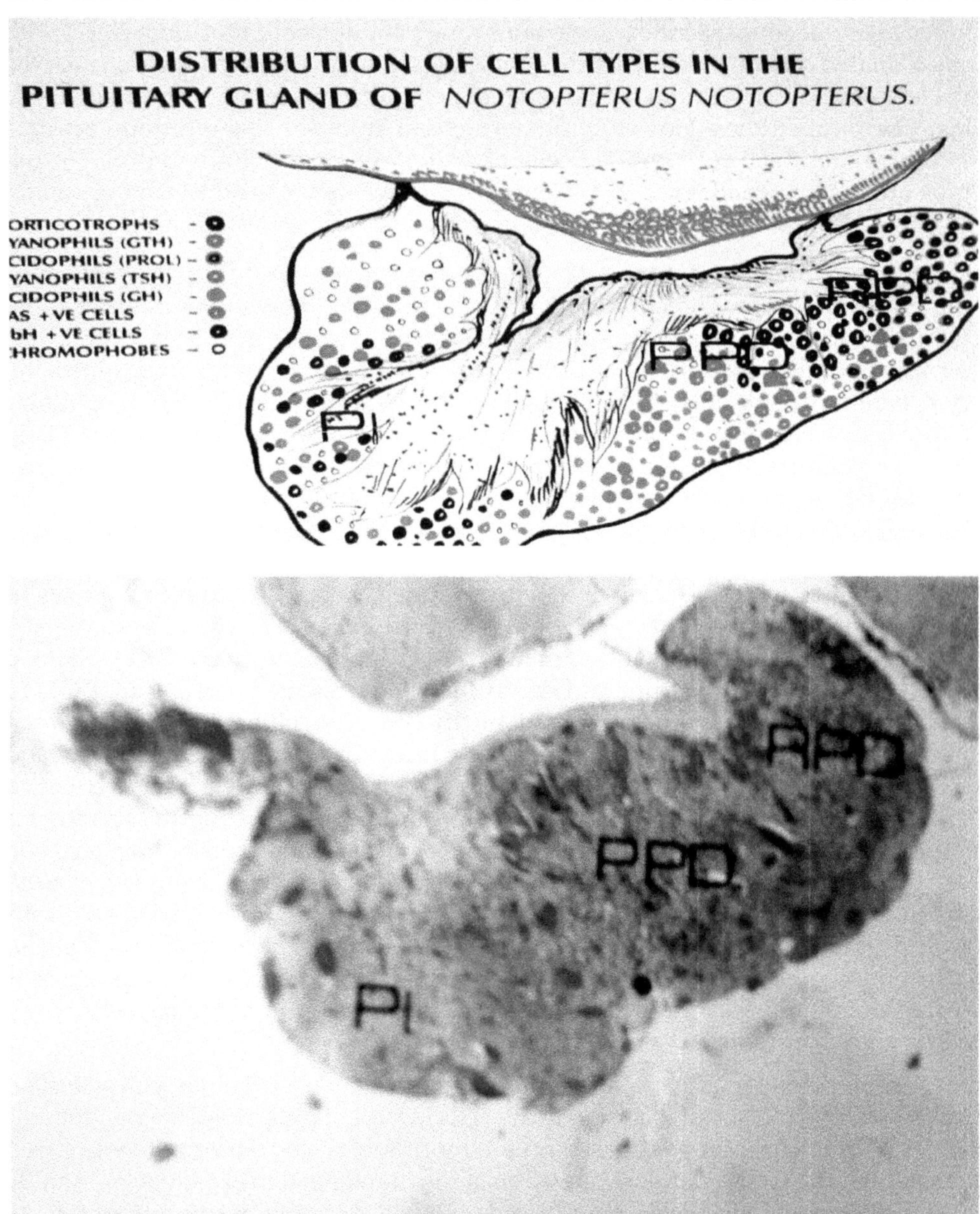

Fig.1.6: Diagramatic presentation of Pituitary cell type distribution in the fish *Notopterus notopterus*

Figure showing pituitary gonadotrophs during breeding period of the fish *Notopterus notopterus* – PAS=PbH+Tetrachrome stains. X1200. Fig: PPD of the pituitary of two age fish showing AF positive cyanophils- AF-Fast green+chrmatotrope 2R-X210, PPD of the pituitary in two age fish showing large number of PAS positive cyanophil cells. PbH+PAS X336, PPD of the pituitary in two age fish showing erythrosine positive and orange-G positive acidophils and aniline blue positive basophils-Cleaveland and Wolf's trichrome -positive. X336, PPD of the pituitary in two age fish showing cell groups of aniline blue positive basophils and erythrosine positive acidophils.X610.

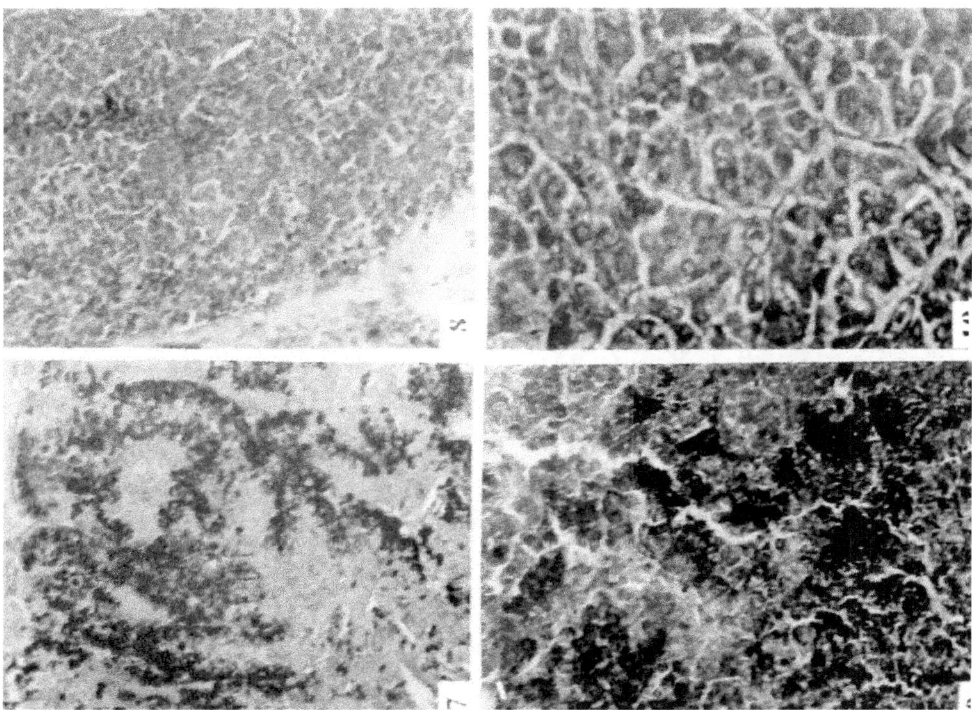

Fig.1.7: PPD of the pituitary showing AF positive cyanophils and large number of PAS positive cells (cyanophils)

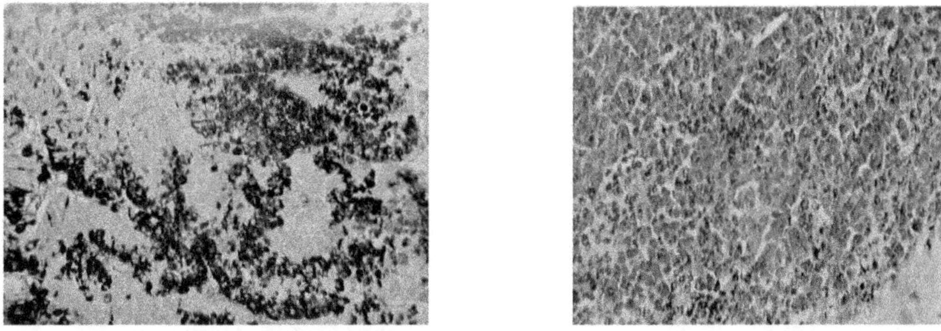

Fig.1.8: Electon micrograph of section of the pituitary in the region of PPD showing GTH cells having different secretary granules associated with blood vessel. X5000.

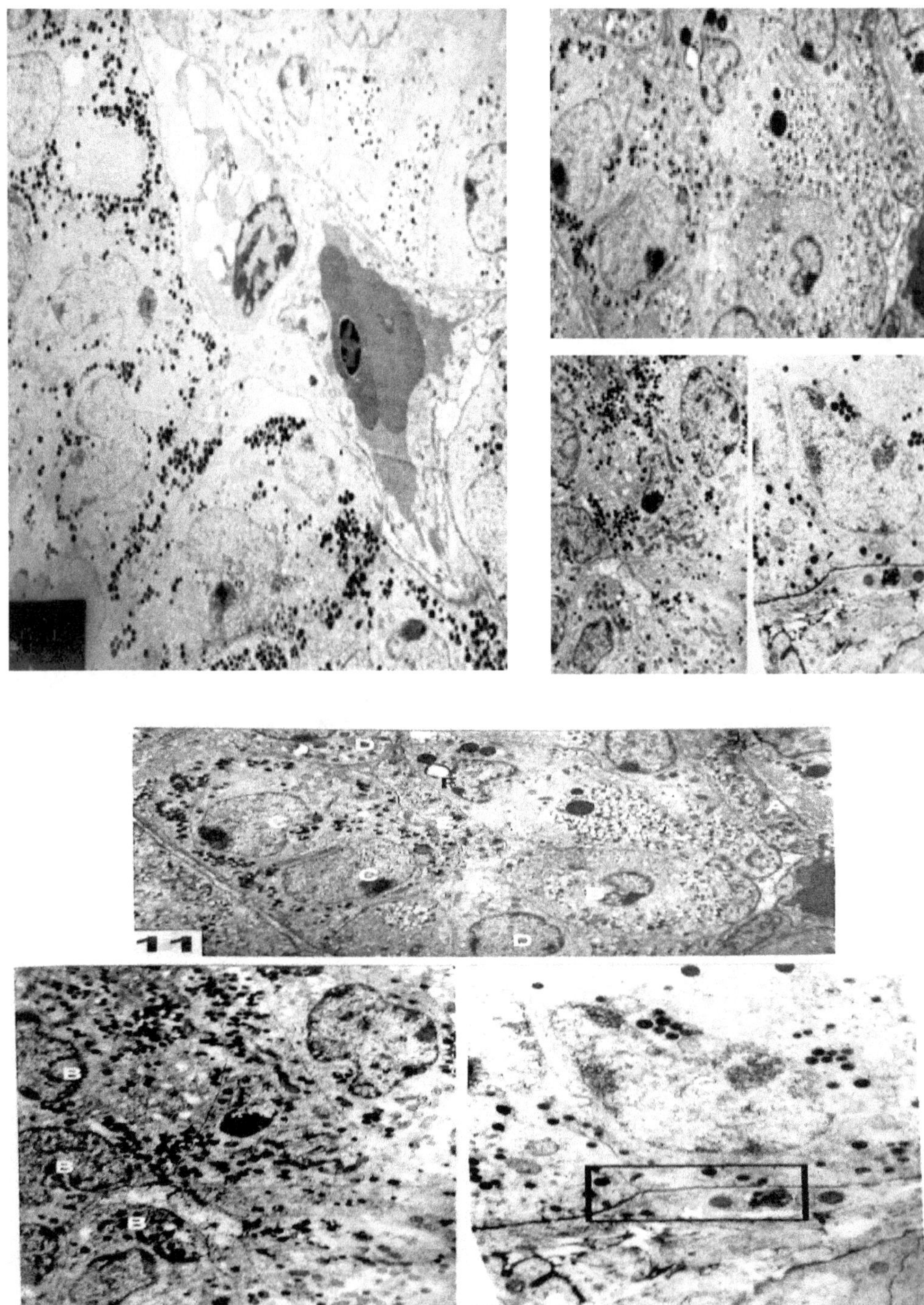

Fig.1.9: Electron micrograph of GTH cells in the PPD having coarse granules, mitochondria (M), endoplasmic reticulum (ER) and glial elements (K), RPD having LTH,STH cells. X6000.

Methodology for Hypothalamo-hypophysial Neurosecretion Studies

Around twenty fish were sacrificed by decapitation and their brains with intact pituitary were fixed as mentioned earlier in Bouin's fluid to study the hypothalamo–hypophysial neurosecretory system by staining in bulk in modified Gabe's aldehyde fuchsin (AF) as used earlier by Sathyanesan (1969). These bulk preparations were used to make thick free hand sections.

About fifty fish were sacrificed by decapitation and their brains with intact pituitary were fixed in Bouin's fluid and Bouin's sublimate.Paraffin sections were prepared at 5 m thickness for the study of hypothalamo – hypophysial neurosecretory system and cytology of the pituitary gland.

The following staining methods were applied for neurosecretion and pituitary cytology:

1. Gabe's aldehyde fuchsin bulk staining for technique modified by Sathyanesan (1969).

2. Aldehyde fuchsin (AF) preceded by acid permanganate oxidation using fast green – chromotrope 2R as the counter stain.

3. Lead haematoxylin (PbH) technique preceded by PAS.

4. Cleveland and Wolfe's trichrome method..

1. Aldehyde thionine (Ath) periodic acid schiff (PAS) orange G (OG) procedure **Gabe's Aldehyde Fuchsin Bulk staining Technique Modified by Sathyanesan (1969)**.

Preparation of Gabe's Aldehyde Fuchsin Stain:

Chemicals

Boiling Water	-	200 ml
Conc. HCl	-	2 ml
Paraldehyde	-	2 ml
Alcohol 70%	-	225 ml
Glacial acetic acid	-	1 ml

Preparation of Stock Solution

Dissolved 1 gm of basic fuchsin in 200 ml of boiling water boiled about 1 minute allowed to cool. Added 2 ml of conc. HCl and 2 ml of paraldehyde and kept in stoppered bottle. Colour was observed every day by putting on filter paper. The red colour was disappeared on 4th day. The solution was filtered, the precipitate in 150 ml of 70% alcohol. This stock solution was kept in refrigerator.

Preparation of Staining Solution

25 ml of stock solution was added to 75 ml of 70% alcohol and 1 ml of glacial acetic acid was added to the solution.

Preparation of Performic acid.

Chemicals

98% formic acid	-	40ml
30% (100 vol.) Hydrogen peroxide	-	4 ml
Conc. H_2SO_4	-	0.5ml

To 40 ml of 98% formic acid added 4 ml of 30% (100 vol.1) H_2O_2 and 0.5 ml of conc. H_2SO_4. Allowed the mixture to stand for at least an hour and used within 24 hours.

Procedure

1. Fixed the tissue in Bouin's fluid for 2-3 days and passed through the 70%, 90% and 100% alcohol.
2. Treated with performic acid till the tissue was settled down (1 to 1 ½ hour).
3. Washed overnight in running water.
4. Kept in 70% alcohol for 1 hour.
5. Stained in Gabe's AF for about for 2 hours.
6. Differentiated 70% alcohol for 12 hours.
7. Dehydrated in 90% and 100% alcohol.
8. Cleared in cedar wood oil or methylsalicylate freehand sections were made. Mounted in DPX.

Gomari's (1951) Aldehyde Fuchsin (AF) Staining Method Preparation of Aldehyde Fuchsin Stain

Chemicals

Basic fuchsin	-	0.75g
70% alcohol	-	100ml
Conc. HCl	-	1.5ml
Paraldehyde	-	1ml

Preparation of Stock Solution

1.5 ml of conc. HCl and 1 ml of paraldehyde mixed to 100 ml of 0.75% basic fuchsin in 70% alcohol. Kept at room temperature until the shade of the mixture darkens to a deep violet colour (about 4 days) shaken intermittent and kept in refrigerator.

Preparation of Staining Solution

1 part of the stock solution added to 3 parts of 70% alcohol.

Preparation of 0.3% potassium permanganate ($KmNO_4$) and 0.3% sulphuric acid H_2SO_4 mixture.

300 mg of $KmNO_4$ was dissolved in 100 ml of distilled water 0.3 ml of H_2SO_4 was added to 100 ml of distilled water and the two solutions were mixed in equal proportions to make 0.3% each.

Chemicals

Potassium Permanganate - 300mg

Sulphuric acid - 0.3ml

Distilled water - 250ml

Preparation to 2.5% Sodium Bisulphate (Na H SO_2) Solution

Chemicals

$NaHSO_2$ - 2.5 g

Distilled water - 100 ml

2.5 g of $NaHSO_2$ was dissolved in 100 ml of distilled water.

Mordant Preparation

Chemicals

Phosphotungstic acid - 4g

Phosphomolybdic acid - 1g

Distilled water - 100ml

4 g phosphotugstic acid and 1 g of phosphomolybdic acid was dissolved in 100 ml of distilled water.

Preparation of Counter Stain (Fast green)

Chemical

Fast green - 0.6g

Distilled water - 100 ml

Glacial acetic acid - 1 ml

0.6 g of fast green was dissolved in 100 ml of distilled water and added 1 ml of glacial acetic acid while using.

Procedure: This method stands for neurosecretory substances in the hypothalamus as well as in the pituitary gland. Fixative used is Bouin's fluid.

1. Deparaffinised 100 sections in two changes of xylene for 5 to 10 minutes and then hydrated by passing through 100%, 90% and 70% alcohol to distilled water.
2. Oxidized one minute with a freshly prepared mixture of 0.3% $KmNO_4$ and 0.3% H_2SO_4 mixture.
3. Rinsed in distilled water (just a dip only) and dicolourised in 2.5% sodium bisulphate (just a dip only) and kept back the solution colour goes off (It take place soon if $NaHSO_2$ is fresh).

4. Washed in running water for 5 minutes and then kept in distilled water.
5. Washed in distilled water since the traces of reducer carried over into the stain may lead to unspecific results (at least it has to be washed twice or more).
6. Stained the sections in aldehyde fuchsin staining solution for 10 to 15 minutes.
7. Rinsed the sections passing through two changes of 95% alcohol and passed on to distilled water.
8. Rinsed the sections briefly in 0.5% of HCl in absolute alcohol.
9. Rinsed the sections in distilled water (kept in distilled water and seen under microscope good slides showing neurosecretory cells, or tracts or tracts or neurohypophysis were selected others were discarded).
10. Slides kept in mordant for 5 minutes.
11. Rinsed rapidly in distilled water.
12. Counter stained for 3 minutes.
13. Rinsed in 95% alcohol with 0.2%acetic acid.
14. Rinsed in 95% alcohol and passed the sections through absolute alcohol 1 and 2 for 3 to 4 minutes in each.
15. Cleared in xylene for 5 to 10 minutes.
16. Mounted in DPX.

Lead Himatoxylin (PbH) Technique, Preceded by PAS Method

Preparation of Periodic acid (0.6%)

Chemicals

Periodic acid	-	0.8 g
Distilled water	-	100 ml
Conc. Nitric acid	-	0.3 ml

0.8 cms of Periodic acid was dissolved in 100 ml of distilled water and then added 0.3 ml of conc. nitric acid.

Preparation of Schiff's Reagent

Chemicals

Basic Fuchsin	-	0.75 g
Distilled water	-	85 ml
Sodium metadisulphite	-	1.9 g
1N HC1	-	15 ml

(1N HCl was prepared by making 8.6 ml of conc. HCl to 100 ml by adding distilled water).

Placed in the bottle with approximate 50-60 ml of free air space, the solution was shaken for 2 hours at intervals of time and left for overnight. 200 mg of activated charcoal was added and shaking was

Carried out every two minutes for 30 minutes and then and then filtered. The reagent was stored in a bottle with minimum air space and kept in dark place (This will decreased the loss of S_2 from the solution).

Preparation of Sodium Metabisulphite Sohition

Chemicals

Lead nitrate	-	2.5 g
Aqueous saturated ammonium acetate	-	50 ml
Haematoxylin Powder	-	2 g
95% alcohol	-	10 m
Tap water	-	140 ml

Solution A: 50 ml of tap water and 2.5 grams of lead nitrate was mixed with 50 ml of aqueous saturated solution so ammonium acetate and then filtered.

Solution B: 2 grams of haematoxylin was dissolved in 10 ml of 95% alcohol. The two solutions were mixed and added 90 ml of tap water Let stand for half an hour and then filtered. The solution was used within 4 to 5 hours.

Procedure

1. Deparaffinised the paraffin sections in xylene and hydrated by Passing through down grades of alcohol from 100%, 90% and 70% kept in distilled water.
2. Treated water periodic acid (0.6%) about 15 minutes.
3. Washed in running water for 15 minutes and kept in distilled water.
4. Treated with Schiff's reagent for 10 minutes.
5. Transferred through sulphate solution (3 changes in each for 1-2 minutes).
6. Washed in running water 5 minutes and kept in distilled water.
7. Stained in lead haematoxylin for 1 hour.
8. Kept in tap water for 30 minutes and kept in distilled water.
9. Dehydrate the sections by passing through 70%, 90% and 100% alcohol.
10. Cleared in xylene and mouted in DPX.

Cleveland and Wolfe's Trichrome (1932) Method

Preparation of Solution A

Chemicals

Erythrosin B	-	1 g

Distilled water - 100 ml

1 gram of erythrosine B was dissolved in 100 ml of distilled water.

Preparation of Solution B

Chemicals

Orange G	-	2 g
Phosphotongstic acid	-	1 g
Distilled water	-	100 ml

1 gram of phosphotongstic acid was dissolved in 100 ml of distilled water. 2 grams of orange G power was dissolved in 100 ml of earlier prepared 1% phosphotongstic acid.

Preparation of Solution C

Chemicals

Aniline blue	-	1 g
Distilled water	-	100 ml

1 gram of Aniline blue was dissolved in 100 ml of distilled water (pH 4 was maintained always).

Proceedure

1. Deparaffinised sections through xylene and passed though down grades of 100%. 90% and 70% alcohol and to water.
2. Rinsed the slides in distilled water.
3. Stained in solution A for 5 minutes.
4. Differentiated in distilled water.
5. Stained in solution B for 2-3 minutes.
6. Drained out the solution B from the slides and without rinsing in distilled water stained in solution C for 5 minutes more the times in thesis solution was washed out the previous stains.
7. Drained out the solution from the slides and soaked the excess stain on blotting paper.
8. Quickly rinsed in 90% alcohol, absolute alcohol and xylene.
9. The fresh change was given with xylene, mounted with DPX.

Methodology for Hypothalamo-Hypophysial Vasculature Studies

For the study of vascularisation of hypothalamo – hypophusial axis, twenty fish wear injected with diluted India ink through the heart and then frozen for two hours. After exposing the root and sides of the brain, the heads were fixed in Bouin's fluid as it is not possible to remove the pituitary intact with brain. They were decalcified by keeping them for longer time in fixative (15 days).

After removing the brain with intact pituitary from cranium, the brains were kept in fresh Bouin's fluid for 24 hours and then transferred to 70% alcohol for 24 hours by giving several changes. The brains were transferred for dehydration by passing through 90% and 100% alcohol. Free hand sections of the brain with intact pituitary were made in the parasagittal, frontal and transverse planes and mounted in DPX

The Structure of Hypothalamo-Hypophysial Axis

The hypothalamo-hypophysical vascularization in *Notopterus notopterus* resembles with that of *Channa punctatus* consisting of arteries contributing to the vascular network of hypothalamus in the region immediately anterodorsal to rostral pars distalis (RPD) which is continued into the neurohypophysis (NH). This network was reported to be comparable to the primary capillary plexus of higher vertebrates. However, they do not form prominent short and long loops as present in mammals but their meshes run into one another with a few portal vessels entering the RPD. The clear capillary plexes and their irrigation in the region of anterior NH forming neuro-adenointerface vasculature.This region has some AF positive neuronal endings and also has neurosecretory tract, suggesting the secretory products might get into the vessels and reach the adenohypophysial cells. This study forms a morphological evidence to demonstrate the presence of hypothalamo-hypophysial portal in this fish. The functional aspect to demonstrate the presence of portal system seen outside the pituitary of *N. notopterus* like that of terapods needs further study. However, the neuro-adeno interface vasculature is prominently seen in *N. notopterus.* Such a interface vasculature has been reported in variety of other Indian fishes.*Clarias batrachus* , *Heteropnustes fossilis* , *Channa punctatus* , *Bagarius bagarius* , *Rita rita. Trichogaster fasciatus. Labeo rohita* , *Puntius sophore. Cirrhinus mrigala* and *Mystus vittatus.* In addition to neuro-adeno interface vasculature tetrapodan type of pituitary portal system has been also demonstrated in these fishes. Recently Bhaskaran and Sathyanesan (1992) have reported a terapod like hypothalamo-hypophysial portal system in the teleost. *Megalops cyprinoides,* in which the complicated loops of the primary capillary plexus are seen on the perikarya of the NLT and the preoptic hypothalomo-neurohypophysial neurosecretory tract. As early as in 1969 Ball and Baker, have reviewed the blood supply to the teleostean pituitary gland and reported that the scattered observations might be interpreted as indicating some kind of portal connection between hypothalamus and adenohypophysis, the important point to emphasize is that in no teleost has any close association between hypothalamic neurosecretory nerve endings and blood capillaries been observed outside the pituitary itself. Thus whether or not these various hypothalamic capillary contributions to pituitary vascularisation represent as it were a foreshadowing (or reminiscence) of the portal condition, there is no reason to regard them as functional teleostean portal systems". Blood from the secondary plexus drains posteriorly into venous system which colasces into one large vein existing in the pituitary. Further, he explains in some species in which vestigial hypothalamohypophysial portal system exists. The teleosts forms a major group amongst the Antenopterigians, the other infraclasses of Actinopterigian group are Chondrostei (*e.g., Polypterus* and *Acipenser*) and Holostei (e.g., *Amia* and *Lepisosteus*), which exhibit a functional blood portal system and does serve to transport neurohormones to the pars system and does serve to transport neurohormones to the par distalis. Perks (1969) in his review observed "the median eminence (ME) and pituitary portal system are common features of the neurohypophysial system from the level of the primitive hagfish up through the vertebrates and the teleosts have developed direct neural contacts with the adenohypophysis and it is possible that the portal system became absolute and was lost". With the available literature on the presence of hypothalamo-hupophysial portal system in some teleost reported by Sathyanesan and his group.

The present study on the vasculature of *N. notopterus* showing capillary plexus at the region anterior to pituitary in the hypothalamus and a prominent neuro-adeno interface vasculature observed is an additional support to demonstrate morphological evidence of presence of a clear vascular link between hypothalamus and hypophysis in *N. notopterus*. To demonstrate the functional aspect of hypothalamo-hypophysical vascular (portal) system in *N. notopterus* needs further analysis and detailed study.

The hypothalamo-neurohypophysial (HN) complex of teleosts comprises of nucleus preopticus (NPO), their axonal tracts, nucleus lateralis tuberis (NLT) and neurohypophysis (NH). Although the teleostean HN complex has been extensively studied and reviewed by several and investigators.The structural details of HN complex has been studied in a few species using *in situ* staining techniques. In some Indian teleostean Sathyanesan and his group has studied the structure of the hypopthalamo-neurohypophysial system using *in situ* staining procedures, describing that the distributional pattern of HN complex remains same, Amongst the Indian cat fishes studied for HN complex are, *Clarias batrachus, Heteropneustes fossilis, Rita rita, Ailia coila, Ompak bimaculaus* and *Mystus vittatus*.

Different species are known to have NPO of different shapes. They may be fusiform, rounded, polygonal and flask shaped. In an extensive review, It has been described the course of the neurosecretory tracts which pass out of the nucleus in an irregular diffuse manner, often leaving in a lateral direction and then bending ventrally before forming a discrete tract in the infundibular floor. In *Nangra punctata* the intact NPO exhibit different shapes when viewed from different planes. Both Pars parvo cellularis (PPC) and pars magnocellularis (PMC) contribute to the formation of right and left main neurosecretory tract which do not enter the pituitary directly, on the other hand they give rise to several pairs of lateral tracts, they join to form a pair of median tract which enters the pituitary, *in situ* studies in *Channa punctatus*, more number of neurosecretory tracts were seen independently running along the lateral aspect and enter the NH In *Micrognathus aculeatus* also these tracts runs laterally but they are loosely consolidated in left and right tracts. In *C. batrachus* The NPO cell appearance varies when viewed from different planes. In ventral view they appear like a paired transverse bands situated on either side of the third ventricle and in cross sectional view the whole NPO is in the form of U shape, in side view, looks like a vertical band. In *Rita rita* The NPO is in the form of two rounded bodies located on either side of the third ventricle and in thick cross sectional view the entire NPO shows that they are in the form of two vertically set elongated bodies. In this species main neurosecretory tracts are composed of loosely set axons for about one third distance towards the hypophysis later they become more compact and give off five or more pairs of lateral tracts which are less prominent in *C. batrachus*. The structure of the HN system of *M. vittatus* is largely comparable to that of other cat fishes described earlier by Sathyanesan and co-workers. In *N. notopterus* NPO cells are oval and oblong shaped when viewed from different planes. Since the shapes of NPO cells are varied and whole NPO extends into the hypothalamus to a varying depths, all its components cannot be seen in one plane, the NPO exhibits like a arch shaped bend in sagittal view. When viewed from the ventral side, the NPO is in the form of rounded structure as in *Rita rita* and it is situated on either side of the third ventricle. The shape of the whole NPO can be also comparable to that of *Nangra punctata*. The NPO cells are having close proximity to ependymal layer as in some cat fishes described.

The third ventricle and ependymal cells of *N. notopterus* has a wide and direct contact with the neurohypophysis. This morphological contact of the ependymal cells with the neurohypophysis and also with the neurosecretory tract has been reported to have some physiological significance. Cerebrospinal fluid and ventricular system have been known as the path way of neuroendocrine integration.

There is a general agreement that AF positive peptidergic fibres originate from the NPO and the negative ones from NLT. It has been suggested that atleast five morphologically distinct fibres innervate adenphypophysis in some fishes. The nerve tracts entering into neurohypophysis of *N. notopterus* gave positive reaction to AF which is originating from NPO. The nucleus lateralis tuberis (NLT) cells are AF negative and extend from the base of the anterior portion of the infundibular base close to the pituitary, midway between optic chaisma and pituitary. The histochemical study of these cells is needed for further analysis. In *N. notopterus* NPO, their axonal tracts entering into the PI along with NLT forms the HN complex. This HN complex of *N. notopterus* is more or less similar to that of *Channa punctatus. It was* observed that although the pituitaries of relatively few of the more then 20,000 species of teleosts have been studied, a wide variation in shape and composition of the gland has been observed. However, it is not the differences, but rather the similarities in structure that are most evident. The teleostean pituitary differs from all other fish pituitaries in that the neurohypophysis.

The experimental studies to locate the different cell types responsible for the production of various hormones are accomplished only in a few species.

It is also reported that there are two types of pituitaries in teleosts, they are platy basic (*e.g.*, found in eeis) and the lepto basic (*e.g.*, found in adult salminids). In platy basic pituitary the neurohypophysis consists of flat floor of infundibulum which sends nerves into adenohypopysis. The adenohypophysis is generally divided into three parts i.e., the rostral pars distalis (RPD), the proximal pars distalis (PPD) and the par intermedia (PI) situated one behind the other, neurosecretory axons penetrate more numerous into the PI than RPD and PPD. In *N. notopterus* the pituitary is platy basic type. It is closely applied to the brain without a distinct stalk, the neurosecretory axons pass through NH and extensively ramifies in the PI. In *N. notopterus* different pituitary cell types were identified on the basis of their response to different staining reactions used in the various techniques. These cells are comparable to that of cells present in other teleostean pituitaries studied by earlier investigators.

The acidophils of the RPD are stainable with orange-G and erthrosin in *N. notopterus.* The erythrosinophilic lactotrophs are some times reported to be chromophobes. In *Tilapia mossambica* , these cells are stainable with orange-G and acid fuchsin as in *Labeo rohita.* In several species prolactin cells are packed with azocaramin or acid fuchsin or erythrosine staining granules. In the fish *N. notopterus* these cells are erythrosinophilic forming a major component of the RPD. On the basis of staining reaction and location as per literature in this species they may be comparable to the prolactin cells of the teleostean pituitary. In *Punctius ticto,* the presence of basophils in predominantly acidophilic RPD. This type of distribution has been clearly noticed in the RPD of pituitary of *N. notopterus.* These basophils are positive to PAS and AF and are found in small numbers of individual cells or in groups.

As the thyrotrophs and gonadotrophs are closely resemble in their staining properties, attempts to separate them purely by their staining reaction have been inconclusive. However, inspite of their similarity they may be structurally different or may be differentiated by their distributional pattern.

In *Anguilla anguilla* ,the TSH cells are located in the RPD and in several species they are located in the PPD. The thyrotophs are also described as spherical cells stainable with analine blue, PAS and AF. In *H. Fossilis* , the thyrotrophs are localised in the middle of the PPD flanked by the gonadotrophs on either sides. The gonadotrophs are known to stain with PAS, AB, AF, ATh and analine blue. In eel and pacific salmon two types of gonadotrophs are reported based on their color differences and granular size, distribution, position and vascularistation. In several teleosts only one type of gonadotrophs are reported. In *N. notopterus* the thyrotophs and gonadotrophs are distributed throughout the PPD. They are positive to PAS, AF and analine blue. These cells are identified on the basis of their staining reactions, but the gonadotrophs seems to be more in number and size and their high staining intensity, when the fish was having mature or maturing gonads. The thyrotrophs has to be irregularly shaped and they are more to the dorsal PPD on the basis of their staining reaction these cells may be comparable to the TSH cells described by earlier investigator.The clear differentiation between thyrotrophs and gonadotrophs needs further experimental study.

Two cells types are generally described in the PI of teleosts and they may secrete two distinct hormones. There is no doubt that this region is the source of MSH , most of the evidences favours the PbH positive cells secretes the MSH. PAS and PbH positive cells are reported in several PI of teleosts. In *N. notopterus* the PI consists of two types of cells, they are PbH positive and weakly stained PAS positive cells. The PbH positive cells are considered as MSH secreting cells in the literature and they are found in all regions of PI in between the neurohypophsial neurons. The PAS positive cells are few and their function is reported in the literature as uncertain. However, it is suggested of calcium metabolism or in the regulation of back ground colour adaptation or in reproduction.

Functional identification of cell types in the pituitary of *N. notopterus* is needed to support the present study carried out in the identification of cell types based on staining response.

Hormones of the Pituitary Gland

The serum content of hormone such as follicle stimulating hormone (FSH) leutinising hormone (LH), prolactin (PRL), thyrotropin, T_3, T_4, estrogen, progesterone and testosterone were estimated by applying radio immunoassay method during four reproductive phases. The hormone assays were carried out at M/s Vijay Hormone Laboratory, Hyderabad. The estimation needed for the hormone assay by RIA were purchased from DPC® diagnostic products corporation, 5700 West 96th Street Los Angeles.

Sample Collection

For the immuno radiometric assay of pituitary hormones (FSH, LH, TSH, PRL) thyroid hormones (T_3 and T_4) and steroid hormones ($E_2 17\beta$, progesterone, testosterone) taken to the laboratory and each fish were sacrified by cutting at the caudal region. The flowing blood was collected in to the test tube and the fishes were dissected to know the sex and accordingly test tubes were marked M for male and F for female. This collected blood is used for further processing.

The amount of blood sample required for the estimation of each hormone is as follows:

The pituitary hormones such as gonadotrophins (follicle stimulating hormone (FSH) and leutinising hormone (LH), prolactin, thyroid hormones such as triiodothyronin (T_3) and thyroxin (T_4), gonadal hormones such as estradiol-17 , progesterone and testosterone has been estimated separately in male and female fishes. The serum concentration of these hormones are presented. The changes in the individual hormones with respect to reproductive phases are presented below:

Pituitary Hormones

a) Follicle Stimulating Hormone (FSH): The serum FSH level (mIU/ml) is determined for entire four reproductive phases. Male FSH level peaked during spawning phase and dropped to a lower level during post spawning phase. There after the level gradually increased during preparatory and prespawning phases. In female fish peak level of the hormone was during prespawning, thereafter gradually reduced reaching lower level during post spawning and elevated during preparatory phase.The increased level of FSH hormone during prespawning and spawning phases indicates their involvement in spermatogenesis and vitellogenesis (folliculogenesis) in male and female fish respectively during these phases of the reproductive cycle.

b) Leutinising Hormone (LH): The serum LH level (mIU/ml) for all the four phases shows that in male fish peak level was during prespawning phase, the level remained more or less similar during spawning and postspawning phases, lower level of the hormone was noticed during preparatory phase. The LH in this fish is involved in steroid production from steroidogenic cells and spermiation process from prespawning to post spawning phases. In females the peak level of the hormone is during spawning phase and lower level in the post spawning phase. The level of the hormone was found to be higher in female in all the phases in comparison to male fishes, thus indicating its involvement in steroidogenesis and spawning activity.

c) Prolactin (PRL): The serum level of prolactin (PRL) in m IU/ml estimated during four phases indicate that the peak level is during spawning phase however, the level of the hormone remained more or less similar in all other phases in males. In females higher level of the hormone is found than males and spawning phase has peak level of the hormone.The hormone level remained more or less similar in other phases. The results suggest that the hormone prolactin is actively involved in spawning activity in both male and female fish in addition to its action in various other physiological activities since the hormone level is found to be constant in other phases, expect increase during spawning phase.

2) Thyroid Hormone (T_3 and T_4): Triiodothyronine (T_3) and tetraiodothyronin (T_4 – Thyroxine) level for both male and female fish during four reproductive phases indicate that T_3 concentration is higher than T_4 in the fish *N. notopterus*. The level of the hormones both T_3 and T_4 are higher during spawning phase and also during preparatory phase in males whereas in females the peak level of the hormones (T_3 and T_4) was higher during prespawning phase with moderately similar levels in other phases.

The increased level of the hormone during preparatory and spawning phases in males and spawning phases in females indicates energy requirement during the phases. Males have higher level of the hormone T_3 than females. The triiodothyronine (T_3) may be the predominant hormone in the fish *N. notopterus*.

Gonadal Steroid Hormones

a) Testosterone: The testosterone level of male *N. notopterus* shows that plasma concentration of (mg/ml) for four reproductive phases. Male fish plasma testosterone levels peaked during spawning phase and dropped to a lower level in post spawning phase. The level of the hormone during preparatory and prespawning phases increased and remained constant during these two phases. Maximum increase is seen only during spawning phase. The plasma level of testosterone in female fish *N. notopterus* though the concentration of the hormone is less in females than males, peak level found to be during spawning phase and remained almost constant in other phases. With minimum level during post spawning phase. The higher level of the hormone testosterone in male fish during three phases suggests its role in spermatogenesis.

b) Estradiol-17β (E-17β): The plasma level of estradiol 17β concentration (pg/ml) values for all four reproductive phases. Plasma estradiol level gradually increased from post spawning to preparatory phase and marginally declined during prespawning and spawning phases in females whereas in males peak estradiol level is during prespawning and thereafter lower level during postspawning. The higher level of estradiol in females during preparatory phase may be because of preparation of female gonad for vitellogenic activity and in male for spermatogenesis.

c) Progesterone: The male and female *N. notopterus* plasma progesterone to concentration (mg/ml) values for the entire four reproductive phases. Female progesterone levels were low during post spawning where after it increased significantly during spawning phase. The plasma concentration of progesterone were a higher in females than males. In male the level of the hormone though less in all the phases significant increase was noticed during postspawning phase. In females the progesterone level increased during prespawning and spawning phase may be because, some developed oocytes have been spawned during prespawning and post ovulating follicles increase resulting in higher production of the hormone.

The plasma levels of various hormones produced from pituitary and other endocrine glands in relation to seasonal gonadal development and maturation have been reported in several species of fish which has proved to be valuable tool in determining the role of hormones in growth of gonads and involvement of associated tissues such as liver and muscle for reproduction in the fish. These studies also provide information in understanding of endocrine control of reproduction in teleosts. The changes in plasma pituitary hormones levels in fish in relation to gonadal activity have been documented in variety of fishes. Idler and Ng, (1983) have reviewed on teleost gonadotropins stating that efforts to equate mammalian with teleost gonadotropins have not been too successful, some teleost gonadotropins have been found to resemble both mammalian FSH and LH in their amino acid compositions and mammalian FSH and LH in chromatographic behaviour. Certain teleost gonadotropins possessed leutinising hormone like biological activities but in other separation of follicle stimulating hormone like and leutinising hormone like biological activities were not clear cut. The FSH and LH were referred to as vitellogenic hormone and maturational hormone respectively in teleost fishes. The pituitary hormone (maturational) content was studied in gold fish at different stages of sexual maturation in the annual reproductive cycle. Further, the variation of plasma titers of the gonadotropin was studied with respect to changes in both reproductive status of the fish and the ambient temperature.

The results show that the pituitary content of maturational hormone increased with the progress of sexual maturation, the plasma level was contingent on both the sexual status and the ambient temperature with gonadal maturation and high temperature favouring a high plasma level. In the fish *N. notopterus* in the present study plasma FSH and LH have been estimated by applying RIA technique using coat A method. The hormones have been quantitatively detected during four reproductive phases. The hormone FSH may be similar to the vitellogenic hormone and LH may be maturational hormone as reported in other fishes. Since these hormones exhibited seasonal rhythms of changes showing gradual increase of FSH from preparatory phase upto spawning phase the time in which active vitellogenesis and spermatogenesis take place. While the hormone LH increase during prespawning and remains more or less same during spawning and post spawning phase in male and higher level of the hormone during spawning phase in female fishes indicates their role in activation of steroidogenic cells and spermiation process in males and steroidogenic and spawning process in females.

Gonadotrophins stimulate the 17β-estradiol production of the ovary of Sarotherodon aureus and androgen production of the testes of *Cyprinus carpio, Gilliethys mirabilus* and *Pleuroneetes platessa.* The SG-G100 induced a free cholesterol depletion from *Channa punctatus* ovarian tissues incubated in vitro. In the fish *N. notopterus* in the present study FSH and LH increases during prespawning and spawning phase, the plasma level of the estradiol and testosterone also increases during these phases indicative of stimulation of gonads for their steroid production. The biochemical changes observed in the liver and muscle during four phase of reproductive cycle may not be the direct effect of these hormones however, they may be acting through the steroid hormones produced from the respective gonads in male and female fish.

Fostier *et al.*, (1983) have reviewed the gonadal steroids in teleosts. A number of workers have reported biosynthesis of steroids from the gonads of variety of fishes and generally steroid hormones such as 17β-estradiol, testosterone and progesterone estimated in the present study in the fish *N. notopterus* have been also identified in number of teleosts during different phases of reproductive cycle. The ovarian steroid hormones in females includes 17β-estradiol, testosterone and progesterone along with other androgens and estrogens. Testicular steroid hormones in males includes testosterone, estradiol 17β and progesterone along with other androgen and estrogens.The steroid hormones such as estradiol 17β, testosterone, and progesterone have been also detected and quantitatively determined by RIA using coat A method in both male and female fish *N. notopterus* in the present investigation during four reproductive phases. Correlation between seasonal changes in plasma levels of gonadal steroids and gonad condition have been well documented in a number of freshwater fish species. Seasonal variation in serum gonadal steroid and associated changes in the biochemical content of the tissues in the present study shows that there existence of correlation in relation to reproductive cycle in the freshwater fish *N. notopterus.* Steroid hormone levels particularly estradiol-17β and testosterone levels were low in postspawning whle they gradually increased during preparatory and prespawning phases in both the sexes. Seasonal variation of steroid hormone level in an intertidal resting fish, plain fish midshipman (a deep water teleost fish) the plasma levels of testosterone and estradiol-17β were determined and found very low throughout the year and peaked during ovaries underwent seasonal recrudescence.

Seasonal changes in serum concentration of the testosterone in males and 17β-estradiol in females and triglycerides and cholesterol in both sexes of *Copoeta copoeta* were determined and it has been observed that seasonal changes in both serum lipids and steroid hormones were associated with reproductive activities.

Estrogen synthesis, mainly 17β-estradiol and/or estrone has been found in most teleosts examined. The estradiol-17β is secreted from the gonads in both male and female fish. In general estradiol is responsible for stimulating vitellogenesis and is also secreted by female gonads. The plasma level of this hormone in *N. notopterus* increases during preparatory and prespawning phases reflects the importance of this hormone. The higher level of this hormone during preparatory phase in female and prespawning phase in male indicates in female its involvement in vitellogenesis and there is also increase in protein content of liver and muscle promoting protein synthesis by this hormone. Estradiol has been reported to stimulate vitellogenesis in teleosts.There is decrease in the level of estradiol during spawning phase compared to preparatory and prespawning phase in *N. notopterus* may possibly indicate that a rapid utilization of the hormone is stimulating vitellogenesis or completion of vitellogenesis as the fish undergoes spawning activity resulting in post ovulatory follicle formation which secrets progesterone. In the fish *Oreochrosmis mossambicus* reported that during resting season there are generally low levels of steroid hormones and during ovarian recrudescence, there is an increase in trophic and steroid hormones and suggests that this relates to vitellogenesis. In the fish *N. notopterus* in the present investigation the pituitary hormones FSH, increase with an subsequent increase in the estradiol production during breeding phase (which includes preparatory, prespawning) reflects its involvement in vitellogenesis. Such observation has been also made in the Indian catfish, *H. fossilis* showing that such estradiol peak corresponds to a rapid vitellogenic growth phase in the oocytes and the level of estradiol maintaining upto spawning phase for protection and to prevent the oocytes from becoming atretic. There is a good correlation between circulating estradiol 17β and protein content of the tissue (liver and muscle) in the present study that the plasma estradiol-17β in *N. notopterus* paralleled increases of the protein content of tissues, there by confirming a role of estradiol-17β in protein synthesis for vitellogenesis. In *N. notopterus* plasma levels of progesterone in female is higher during spawning phase and lower during post spawning and preparatory phases. This reflects formation of post ovulatory follicles and increase in progesterone level during spawning and also some increase during prespawning phase indicates that some developed oocytes have spawned.

An inverse relationship was found to be seen between plasma level of progesterone and estradiol-17β levels in the female fish *N. notopterus,* when plasma progesterone levels are high, estradiol levels are low. This observation indicates the development of ovarian follicles and recrudescence. Such observation has been noticed in the fish *O. mossambicus.* In the male fish earlier studies have shown that increasing plasma levels of progestin with sperm production and androgen synthesis and stimulation of mitosis of germ cells concerning estrogens, biosynthesis *in vitro* was also reported in the testis of ambisexual species ,in the male fish *N. notopterus* plasma levels of progesterone was very low compared to estradiol. However, progesterone increases during post spawning phase and preparatory phase may indicate its involvement in the testicular activities associated with probably androgen synthesis and stimulation of germ cells.

The plasma testosterone levels in *N. notopterus* found to be increased gradually from preparatory phase to spawning phase with decline during post spawning phase indicates its role in gonadal growth particularly for spermatogenesis in male and vitellogenesis in female.

The testosterone levels are higher in males than females corresponds to its requirement for spermatogenesis in males than females. The general increase in testosterone from preparatory phase to spawning phase indicate the commencement of spermatogenesis within the testis, such increase level of testosterone in both male and female fish has been reported in the fish, *O. mossambicus* and suggested its need for spermatogenesis in males and growth of gonads in females. Earlier studies also show that the androgen production from the ovaries has been reported in fish species *in vitro* studies. In the grey mullet, *Mugil cephalus,* the ovarian production of the ketotestosterone increases with the development of vitellogenesis and then decreases after spawning.

Prolactin has been shown to be involved in lipid metabolism and fat storage and also to reduce thyroxine levels of the serum in fish. In teleosts prolactin is known to control a wide spectrum of physiological processes such as osmoregeulation, growth, metabolism and reproduction. The prolactin level recorded in the present study during four stages of reproductive cycle indicate that the hormone level increases during spawning phase. However, the level remains more or less uniform in other phases in both male and female fish *N. notopterus*. Female fish prolactin level is higher than male fish prolactin.

Experimentally prolactin has been shown to be involved in ovarian steroidogenesis of the guppy *Poecilia reticulate* particularly regulating estrogen synthesis to facilitate rapid vitellogenesis i.e., regulation of prolactin for vitellogenin uptake by the oocytes. The increasing level of prolactin on approach to, breeding period (prespawning and spawning phases) in the fish *N. notopterus* also indicates its role in reproduction in this fish specially for vitellogenesis in female fishes. Ovine prolactin also induced a steroidogenic response in non hypophysectamised fish.

The higher level of prolactin in *N. notopterus* during breeding phase may also be involved in gonadal steroid production as the gonadal steroid hormones both estradiol 17β and testosterone in both male and female increases during these phases. Matty, (1985) has discussed the role of thyroid in teleost reproduction. Seasonal thyroid hormone levels have been measured in salmonids. In the trout *Salmo gairdneri*. There appears to be two peaks in plasma concentration , while in lake Ontario cohosalmon, there is but one peak of serum T4 and T3. It has been further suggested that two thyroid cycles occur in teleost fish, one concerned with temperature compensating mechanism and the other with reproduction. The thyroid hormones (T3 and T4) measured in *N. notopterus* in the present study indicates that T3 level is more than T4 level in both male and female fish and T3 level is more than T4 level. The changes in the level of T3 and T4 during four reproductive phase indicates that the level is almost uniform in all the phases with an higher level during spawning phase in males and during prespawning phase in females. The fish *N. notopterus* comes to the surface of the water regularly for gulping air and in constant movement, needs higher metabolic rate which may be the reason why the thyroid hormone found to have constant level in all the four phases. The increase in the level of the hormone during prespawning and spawning phase may be related to still higher metabolic rate for active reproductive activities during these phases.

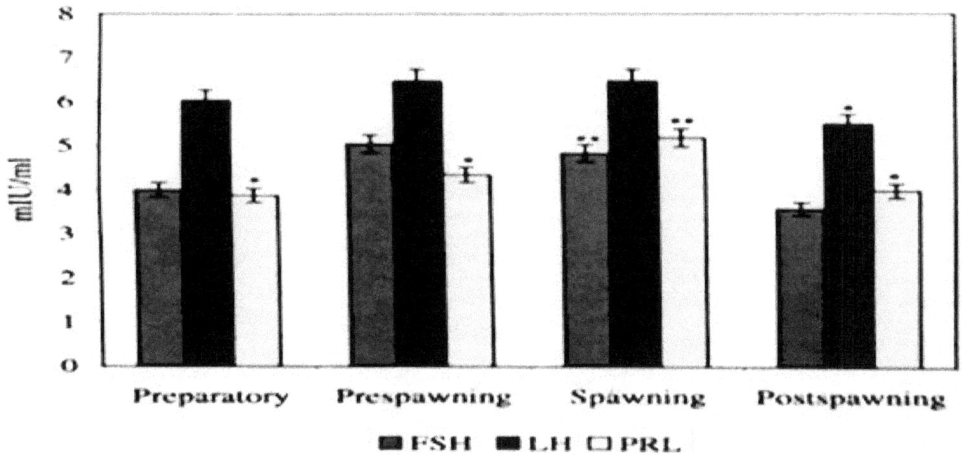

Fig.1.9: Some pituitary hormones in the female fresh water fish, *Notopterus notopterus* during different reproductive phases

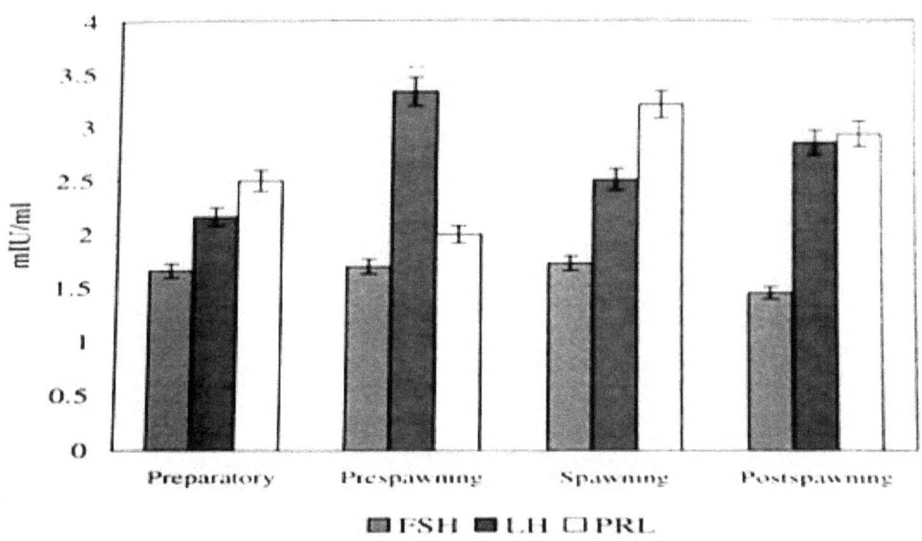

Fig.1.10: Some pituitary hormones in the male fresh water fish, *Notopterus notopterus* during different reproductive phases

Chapter-2

Structural Organisation of Pineal Gland and Melatonin

In most of the poikilothermic vertebrates the pineal complex has two components. In fish they are the pineal and parapineal organs which look similar during ontogeny but reach different levels of differentiation in the adult. The parapineal organ remains more or less rudimentary while the pineal organ grows to form a relatively large vesicle located dorsally to the fore brain immedietly below or within the skull roof. The pineal if often differentiated into a proximal slender pineal stalk and distal expanded end vesicle. The wall of the pineal organ is formed by unstratified epithelium which is strongly folded and may almost obliterate the central lumen of the pineal organ.The pineal epithelium of teleosts consists of photoreceptor cell, neurons and ependymal cells which are often called supportive cells, interstitial cells and some glial cells. Some oligodendrocytes have been found surrounding neural axons and macrophages are found in the central lumen of the pineal.The photoreceptor cells produce the endolamine, melatonin. The pineal gland translates photoperiodic information to a hormonal signal which would then serve as a messenger to every organ of the body. The size of the pineal organ varies considerably amongst fish species. In some species pineal gland is large and covers most of the telecephalic (e.g. pike or salmon) and other species such as cod, sea bass and tilapia the pineal is less conspicuous and lacks its charectoristic vesicle form. The pineal gland (also called the pineal body, epiphysis cerebri, epiphysis conarium or third eye) is a small endocrine gland in the vertebrate brain; it produces the serotonin a derivative of melatonin a hormone that effects the modulation of wake/sleep pattern and seasonal function. The pineal complex of teleostean fish consists of pineal and parapineal organ and this pineal organ is directly a photosensory organ that contains photoreceptor cells similar to those of the retina. In some teleosts, the area of the skull covering the pineal organ is specialised as "the pineal window" this window seems to facilitate light penetration to the organ. In most of teleosts, the photoreceptor cells exhibit well developed outer segment and form ribbon-containing terminals on secondary neurons,whose axons contribute to the pineal tract connecting the pineal organ with brain stem and pineal tract fibers run to the hebenular nuclei, sub commissural organ and pretectum.The parapineal organ of most teleost (*eg. Phoxinus phoxinus, Anguilla anguilla*) is a small round body located on the left side of the brain above the epithalamus and tract organinating from the neurons of the organ connect with the habenular nucleus.The organ contains neurons, glial cells and parapinealocytes of the photoreceptor type ,their outer segments bind rhodopsin antesera indicating the presence of rhodopsin or porphyropsin and light percepting ability. The available ultrastructural studies of teleost pineal organ there appear to be large interspecific variation with regard to fine structure of pineal photoreceptor cells.

The pineal gland of the fresh water fish *Notopterus notopterus* was studied by fixing brain with intact cranium in the Bouins fluid for 72 hours and dehydrated in ascending grades of alcohol, cleared in toluene and embedded in paraffin wax. the free hand sections were taken of the block and the sections were stained in haematoxyline and Eosine stains and observed under compound microscope. Based on the observation a free hand sections of brain along with cranium, the diagrammatic presentation has been constructed.

A well vascularised pineal complex of this species consists of three saperate but distinctly connected components has been noticed.The pineal complex present just below the roof of the cranium has end vesicle, a long and curved pineal stalk(PS) and a dorsal sac (DS).furthermore the base of pineal gland is present just beside the posterior commissure. Histological observation of the pineal gland consists of about 80% of cells that have rounded vesicular nuclei, these are pinealocyte cells positively stained with haematoxylin and eosine. The remaining cells are astrocytes which cannot be seen on H&E slides.

The available ultra structural studies of teleost pineal organ appear to be large interspecific variation with regard to fine structure of pineal photoreceptor cells. The pineal gland develops from the posterior of the two diancephalic evagination, the pineal organ is recognised with stalk and an end vesicle, pineal window is used for the translucent part of the cranium above the pineal organ which admits light. Although the shape and size of end vesicle varies widely depending on the species, it may be larger and conspicuous or insignificantly reduced and rudimentary.In the fish, *Catla catla* the pineal complex consists of three saperate but distinctly connected components such as an end visicle (EV), long pineal stalk (PS) and dorsal sac. The epithelial cells lining of end vesicle consists of cells which have rounded vesicular nuclei and long cytoplasmic processes that reach the lumen and suggested that they are photoreceptor cells. The cells of the pineal stalk have some similarity with those of the end vesicle, while dorsal sac cells appear columnar and ciliated. The morphology of pineal complex in the fish, *Notopterus notopterus* consists of distinctly connected end vesicle, long pineal stalk and a dorsal sac as reported in the fish, *Catla catla*. The cytology of the end vesicle consists of cells which might be pinealocytes which have round vesicular nuclei, positively stained with H and E. histologically the teleostean pineal closely resembles a sensory structure. The pineal organ is made up of three kinds of cells photoreceptor cells, supportive cells and ganglions, the first two cells can be recognised and the third one is less prominent due to their very little number. The supportive cells present in between photoreceptor cells stain less intensely and have numerous blood vessels surrounding these cells.The photoreceptor cells are the most numerous of epithelial cells. In the fish, *Notopterus notopterus,* the cells are which were observed in the end vesicle are probably photoreceptor cells. The pineal stalk of the fish, *Notopterus notopterus* is long and curved, basally attached to dorsal sac which is distinctly present.

Melatonin is important in controlling the seasonality of reproduction in fish by stimulating the final stages sexual maturation. The control of reproduction by melatonin is species specific, the timing of reproduction in fish can be affected by photoperiod in conjuction with clear diel and seasonal pattern of melatonin. Many experimental studies have shown that melatonin has a role in reproduction. The physiological significance of melatonin in the regulation of testicular events has been studied in a major Indian carp *Catla catla* and was evaluated through studies on the effect of graded doses of melatonin exogenously administered for different durations. The testicular response to exogenous melatonin in each reproductive phase was almost identical. The testicular function in both melatonin treated fish inhibited during prespawning and spawning phases and concluded that melatonin plays a significant role in the regulation of annual testicular event in a subtropical surface dwelling carp, *Catla catla.* However, the influence of this pineal hormone in the seasonal activity of the testis varies in relation to the reproduction status of concerned fish.

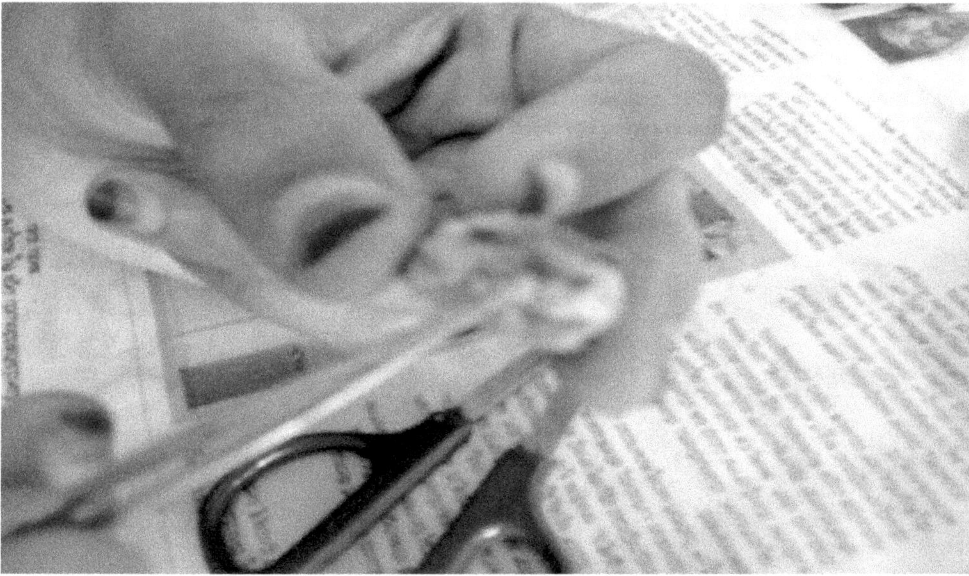

Dissection of cranium in the fresh water fish, *N.notopterus* for the location of pineal complex.

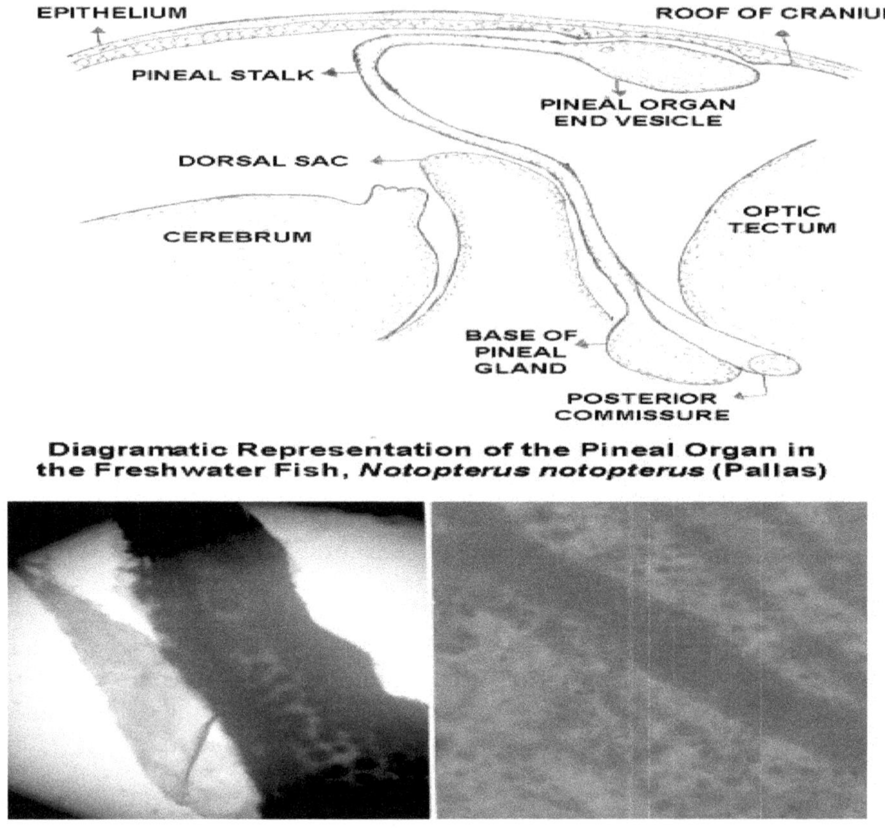

Fig.2.1: Diagramatic presentation of pineal complex in the fresh water fish, *Notopterus notopterus* and photographs showing epithelial cells (Pinealocyte) in the end vesicle.

Chapter-3

Structure and Distribution of Thyroid Gland

The thyroid gland has been studied extensively studied in teleost fishes. The thyroid of most teleosts consists of isolated follicles scattered around the ventral aorta and its branches inside the gills. This pattern of scattered follicles can extend quite widely outside the pharyngeal region, in some fishes, example-In platyfish, follicles are found in the eye, kidney and other organs. This migration of so called heterotropic follicles, from the pharyngeal region is probably due to the fact that the gland is not encapsulated and surrounded by connective tissue. The presence of functional thyroid follicles in the head kidney of some Indian teleosts by us (Kulkarni and Sthyanesan, 1978). In variety of vertebrates, the thyroid plays very important role in growth, development and sexual maturation. It has been reported that, there appears in bony fish also to be a close correlation of thyroid and reproductive activity. Although cause and effect are often difficult to establish. Furter, the making of teleost hypothyroid either by treatment with goitrogens or radiothyroidectomy or surgical thyroidectomy does not invariably retard gonadal development and certainly goitrogens effects may be regarded as toxic menifestaions of rapidly dividing tissues. Thiourea has been used as one the chemical for thyroidal studies in fishes to understand relation ship of the thyroid with reproduction. The inhibition of thyroid activity in fishes by the administration of chemicals such as sulphonamides and thiourea has been reported and these chemical substances prevents the iodination of the tyrosine. However accumulation of iodine takes place.

The observation of thyroid follicles distribution in the fish *Notopterus notopterus,* the lower jaw of the fish was fixed in the Bouins fluid containing Picric acid (75 ml), Glacial acetic aciid (10 ml) and Formalin (20 ml). The tissue was kept in Bouins fluid for a period of one month for dissolving cartilaginous part. The Bouins fixed tissue were dehydrated in different grades of alcohol, cleared in toluene and embedded in the paraffin (melting point of paraffin about 58-62 degree centegrade).

Serial sections were prepared at 5-6 micron thickness on rotary microtome. The sections were placed on slides and dried. The sections were deparfinised by treating with toluene and brought to distilled water by hydration placing in different grades of alcohol and stained in Haematoxyline and eosine. Some sections also stained in PAS (Periodic acid Sciff) stain to observe staining intensity indicating glycogen concentration in the thyroid follicular lumen. To verify weather thyroid of fish,Notopterus notopterus, the fish were treated on exposure with thiourea at a concentration of 0.03 % and maintained for a period of three months and every one month the fish were sacrificed for the observation of thyroidal follicles status. The gonads such as testis and ovary were also fixed for observation.

The thyroid follicles in the fish, *Notopterus notopterus* are distributed just by the side of ventral aorta in the pharyngeal region.The follicular cells are squamous and the colloid the lumen of the follicles was eosinopilic and PAS posoitive.After the exposure of thiourea, the thyroid follicular cells hypertrophied and colloid in the follicles exhibited vacuolation. After three months of complete exposure the thyroidal follicular cells were extensively hypertrophied and in some of the follicles the lumen reduced markedly without any colloid. The thiourea exposure affected the gonadal condition, the testicular lobules were disorganised exhibiting different stages of degeneration. Similarly, the ovary also exhibited degenerative condition with majority of ovarian follicles became atretic.

In teleost fishes thyroid gland hormones are known to influence the reproductive process. The seasonal changes of thyroid histology has been correlated with gonadal development during different phases of the reproductive cycle, suggesting that there is influence of thyroid hormones in gonadal development. In fishes thiourea and propyl thiouracil are reported to inhibit thyroxine secretion and low levels of circulating hormones activated the thyrotrophs of the pituitary gland to produce more of thyroid stimulating hormone, as a result hypertrophy of the thyrotrophs of the pituitary gland. There are two kinds of reports on the action of thioure on fish gonads as direct and indirect action of thiourea on gonads. in the indirect action of thiourea on fish gonads , the decreased thyroid activity in response to antithyroid drugs, which accelerated TSH production and there was a simultaneous reduction in gonadotrophin concentration and in turn retardation of gonadal activity.

Thyroid Hormones and Its Effects

Thyroid hormones (T_3 and ,T_4): Triiodothyronine (T_3) and tetraiodothyronin (T_4 – Thyroxine) level determined for both male and female fish during four reproductive phases indicate that T_3 concentration is higher than T_4 in the fish *N. notopterus*. The level of the hormones both T_3 and T_4 are higher during spawning phase and also during preparatory phase in males whereas in females the peak level of the hormones (T_3 and T_4) were higher during prespawning phase with moderately similar levels in other phases. The increased level of the hormone during preparatory and spawning phases in males and spawning phases in females indicates energy requirement during the phases. Males have higher level of the hormone T_3 than females. The triiodothyronine (T_3) may be the predominant hormone in the fish *N. notopterus*. Matty, (1985) has discussed the role of thyroid in teleost reproduction. Seasonal thyroid hormone levels have been measured in salmonids. In the trout

Salmo gairdneri. There appears to be two peaks in plasma concentration , while in lake Ontario cohosalmon, there is but one peak of serum T_4 and T_3. It has been further suggested that two thyroid cycles occur in teleost fish, one concerned with temperature compensating mechanism and the other with reproduction. The thyroid hormones (T_3 and T_4) measured in *N. notopterus* in the present study indicates that T_3 level is more than T_4 level in both male and female fish and T_3 level is more than T_4 level. The changes in the level of T_3 and T_4 during four reproductive phase indicates that the level is almost uniform in all the phases with an higher level during spawning phase in males and during prespawning phase in females. The fish *N. notopterus* comes to the surface of the water regularly for gulping air and in constant movement, needs higher metabolic rate which may be the reason why the thyroid hormone found to have constant level in all the four phases. The increase in the level of the hormone during prespawning and spawning phase may be related to still higher metabolic rate for active reproductive activities during these phases. When biochemical changes of protein, glycogen, lipid and cholesterol in hepatic and muscle tissue are correlated it indicates that there is a reduction of these biochemical contents in response to increase in the thyroid hormones.

The thyroxine harmone tablets of 10mg were implanted in the muscle by making a small incision in the dorsal region. The trade name for thyroxine is Eltroxn, a synthetic thyroid hormone. After the implantation the incised area was covered with vasoline and left for 10 days. The fishes were divided in to two groups of 10 each. Group I served as control which was not received thyroxine hormone; Group II served as experimental one which received 10mg of the thyroxine hormone tablet implantation. This experiment was carried out during the prespawning phase. After 10 days treatment all the experimental, and control fishes were sacrifised by decapitation and tissues are analysed for biochemical contents. The biochemical contents were also estimated in the fish implemented with thyroxine tablets during prespawning phase. The results indicates that all the hepatic and muscle protein, glycogen, lipid and cholesterol contents were less in thyroxine treated fish as compared to control.

During differenent reproductive phases.

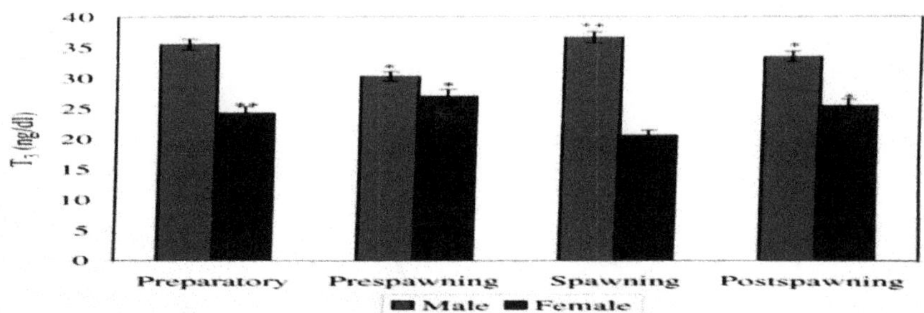

Fig.3.1: Serum thyroid hormone T3 in the male and female fresh water fish, *N.notopterus*

During differenent reproductive phases.

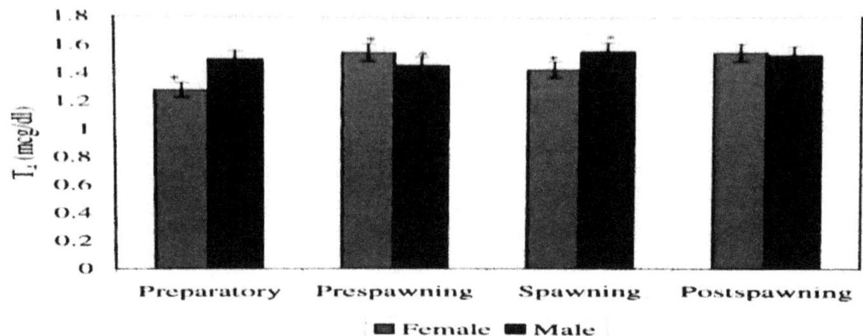

Fig.3.2: Serum thyroid hormone T4 in the male and female fresh water fish, *N.notopterus*

Chapter-4

Structural Organisation of Adrenal Components and Corpuscles of Stannius

Structural Organisation of Adrenal Components and Corpuscles of Stannius

In teleost fish,the adrenal tissue is situated in the lymphoidal pronephric head kidney around the post cardinal vein. The histology and the distributional pattern of the cortical or interregnal and medullary or chromaffin cells have been extensively investigated in several species of teleosts. Kulkarni and Sathyanesan and others have studied the histochemical and histoenzymological response of adrenal component including enzyme localization of 3 hydroxysteroid dehydrogenase in some Indian teleost fishes, which indicates steroidogenesis in the adrenocortical components. The chromaffin identification and their distribution amongst the adrenocortical cells have been also studied by applying potassium dichromate reaction. The adrenals of some fishes reported are known to produce corticosteroids and catecholamines. In the fish, *Notopterus notopterus* these adrenal components are not distinctive and are closely mixed amongst the haemopoitec tissue. Recently in the same fish the adrenal components consisting of cortical and medullary cells by applying various histochemical techniques has been reported by Padmanabh Chakrabarti (2014) that the adrenal gland of *Notopterus notopterus* (Pallas) consisted of interrenal and chromaffin cells, distributed around the main branches of posterior cardinal vein in most anterior part of the pronephric kidney. Seasonal histological characteristics of the interrenal and chromaffin cells in relation to testicular activity of *N. notopterus* were studied by adopting different staining techniques. The cytophysiological features of the interrenal, chromaffin cells and various spermatogonial cells were observed during different reproductive phases. Histologically, the adrenal gland of *N. notopterus* consists of interrenal and chromaffin cells which are separated from each other and distributed around the main branches of posterior cardinal vein in the anterior kidney. The interrenal and chromaffin cells are surrounded by plasma membrane, which separates the cell from adjacent cells. The cytoplasm of the interrenal cells shows a basophilic colouration with Mallory's triple stain and haematoxylin-eosin stain. The chromaffin cells are eosinophobic and round in shape. The interrenal and chromaffin cells are characterized with centrally placed spherical conspicuous nuclei.

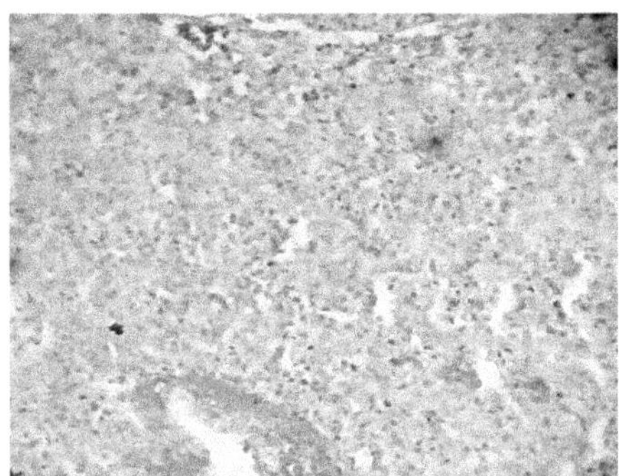

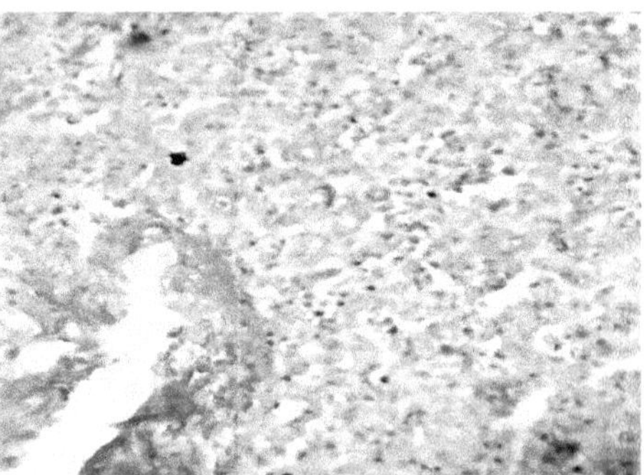

Fig.4.1: Section of head kidney containing adrnocortical and chromaffin cells in the fish *Notopterus notopterus*

Adrenal Steroids and their Effects

The interrenal (adreno cortical) cells are the source for cortisol hormone in fish and is known to be involved in the general homeostasis of energy as well as in acute and chronic responses to stress. The hormone cortisol is the principle corticosreroid in teleost fishes and its plasma concentration rise dramatically during stress (Mommsen et al;1999). One of the most commonly measured indicators of stress in fish is the concentration of the major circulating corticosteroid,cortisol and this hormone can be measured easily and accurately using commercially availableioimmunoassay technique (RIA). There is increasing evidence indicating that cortisol directly and/or indirectly plays an important role in intermediary metabolism in addition to its role in other functions. Acording to Mommsen et al; (1999) plasma cortisol measurement was used as an indicator of stress and few studies have actually measured the kinetics of cortisol in fishes. In view of cortisol having multifaceted role in both metabolically and physiologically the following studies were undertaken in the fresh water fish, *Notopterus notopterus*- 1.Changes in the serum level of cortisol during a day was studied in the fish, *N.notopterus*, 2. Changes in the level of the hormone during different reproductive phases estimating serum cortisol in both male and female fish saperately for four phases, 3. The cortisol was injected during different phases of the reproductive cycle and the serum hormone level was measured.

The plasma cortisol level was studied in the fresh water fish, *Notopterus notopterus* by applying RIA method and a comparison was observed in male and female fish. The level of the hormone ranged between 11µ/ dl to 25µ/ dl for the male fish and 5µ/dl to 32µ/dl for female fish during a single day. The serum cortisol level was also studied during different timings (at 8.30 am, 13.30 and 20.30 in a day) indicated that during morning hours cortisol level was low while in the after noon ang evening hours the cortisol level seems to be higher in both male and female fishes. Thus it indicates that the fish feeds during morning hours and since metabolic activity increases during afternoon hours resulting in the increase of serum cortisol level. It was observed that the fish is active upto evening hours till night,coming to the surface for gulping air and moving continuously. The reproductive cycle of *N. notopterus* was studied. The cortisol level studied during four phases of the reproductive cycle in *Notopterus notopterus* indicated that the cortisol increases during preparatory and spawning phases whereas reduces in other phases (post spawning and resing). The increase level of cortisol during preparatory and spawning phases relates its involvement in the reproductive activity as the fish needs increased metabolism (energy requirement) for spermatogenesis and vitellogenesis and subsequent spawning of gamates. Seasonal effect of cortisol on body weight and condition factors in male and female fish, *Notopterus notopterus* during four reproductive phases indicated that the cortisol hormone has stimulating effect on the body weight in different phases specially in the post spawning fish ,it is possible that it may promote active feeding and enhanced body weight during the treatment of cortisol. The condition of the fish was also found to be improved during the treatment of cortisol hormone. Effect of cortisol on the gonads and liver in both male and female fish, the gonadosomatic and hepatosomatic index increased.The young oocytes transformed into vitellogenic oocytes in all the phases after cortisol treatment. The biochemical contents such as cholesterol , protein and lipid found to be altered depending upon the requirement for ovarian growth. Although in male fish the cortisol had negative effect during mature phase,it has positive effect during immature phase.

The appearance of different spermatogenic stages during immature phase indicating activation of testis for spermatogenesis. The increase in cholesterol and lipid content after cortisol treatment during post-spawning phase are all supportive to positive action of cortisol in the fish. The hepatocytes and the biochemical contents changes in response to hormone treatment are in positive trend of action for growth and gametogenesis indicates that cortisol induced increased ovarian and testicular activity may be due to increase in the metabolic activity through the involvement of hepatic cells specially during non-breeding period providing support for cortisol role in energy production and its utilisation for reproductive activity in the fresh water fish, *Notopterus notopterus.*

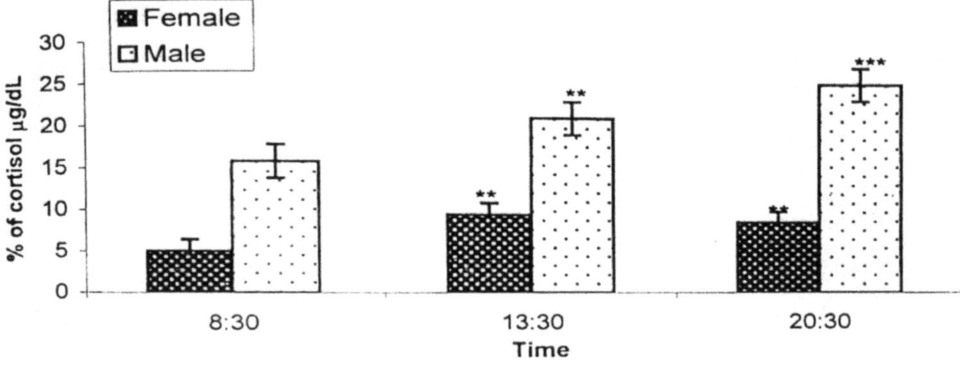

Fig.4.2: Serum cortisol in the fish, *Notopterus notopterus* at different time of the day and seasonal changes.

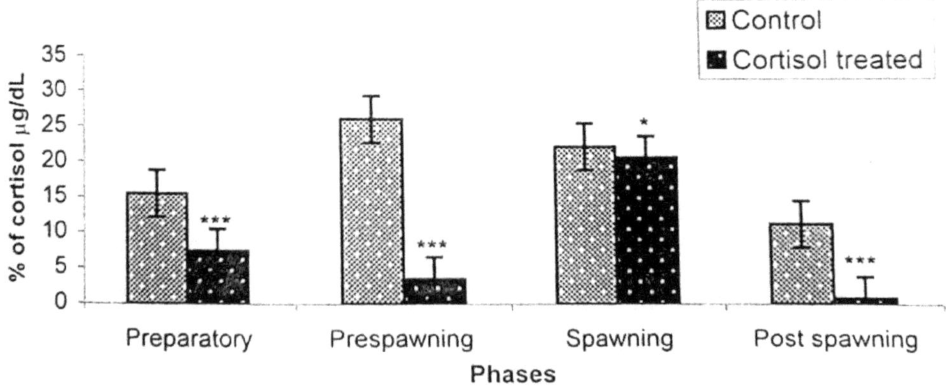

Fig.4.3: Serum level of cortisol hormone during different phases of reproductive cycle of male fish, *N. notopterus*

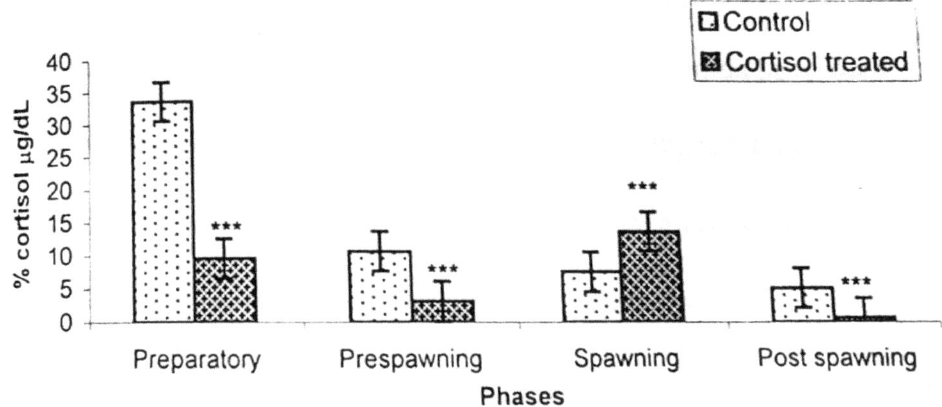

Fig.4.4: Serum level of cortisol hormone during different phases of reproductive cycle of female fish, *N. notopterus*

Adernal Catecholamines

The medullary cells (also called chromaffin cells) of the fish give positive reaction to dichromate, iodate reactions exhibiting yellow to brown colouration resulting in differentiation of two different types of cells as adrenalin and noradrenalin secreting cells. These cells are sources of catecholamines, adrenalin and noradrenalin which are biochemically shown to be present in the fishes studied.The ratio of adrenalin to noradrenalin secreted by the adrenal medullary cells varies with the species. In the eel fish the plasma adrenalin and noradrenalin was estimated to be 1.3 to 3.4 pmol/ml. In gasterosteaus aculatus fish, the two hormones found to be equal. The chromaffin cells in the fish, *Notopterus notopterus* are distributed amongst the adrnaocortical cell and are in large numbers.

The general effects of adreanalin in fish through administration has been studied,it causes hyperglycemia and increases the plasma fatty acids and hyperlipidimia. Adrenalin was reported to have an inhibitory effect on metabolic rate in fishes, it reduces oxygen consumption in *Tilapia mossambica.* Adrenalin also stimulates glycogenolysis in liver and muscle and increases oxygen consumption along with heat production. The adrenalin was injected into another fish by the author (*Channa punctatus*) during non breeding period of the reproductive cycle indicated that there is increase in the body weight,increases gonadosomatic index (GSI), reduces hepatosomatic index (HSI), hepatic nuclear hypertrophy and reduction in the hepatic biochemical contents along with positive changes in the ovarian and testicular histology indicating involvement of the hormone for reproductive activity which may be for gearing up of metabolism to provide large amount of energy necessary for protein and aminoacid synthesis. The reports concerning the role of adrenal catecholamines in fish reproduction indicate that the adrenalin found to be elevated during breeding period and remain at higher level during breeding period reflects necessity of the hormone for breeding purpose.

Structural Organisation of Corpuscles of Stannius

The general organization of the corpuscles of Stannius in *Notopterus notopterus* shows that, it has a spherical structure and are found lying on or embedded in the anterior portion of posterior part of the kidney (Fig.4.2). Only one pair of the corpuscles was observed (Fig.4.3). The gland is highly vascularised with rich innervation. The diameter of the corpuscles ranges from 0.5-2.0 mm which varies during different phases of the reproductive cycle. The corpuscles of Stannius of *Notopters notopterus* are whitish (Fig.4.4) and have a fibrous capsule and are composed of short columnar cells, closely packed in columnar groups.

Histology of Corpuscles of Stannius

The corpuscles of *N. notopterus* spherical in shape, the cells of the gland are usually arranged in strands, separated by septa of connective tissue continuous with outer fibrous capsule. The septa contain all the vascular and nervous elements. The vascularization of the Stannius corpuscles is rich, which results in a better access of the gland cells to the blood circulation. The capillaries are associated with renal arteries and cardinal veins. The Stannius corpuscles probably also have a rich nervous supply. The nerves are found in close association to the corpuscles. This indicates rich innervations.

The cytoarchitecture of the corpuscles of Stannius (CS) in *Notopterus notopterus* shows that four types of arrangement of connective septa and cells. The first type which is common in which cells are arranged in the form of cords lined by single layer of cells, along thin septa exhibiting circular appearance. In the second type, a thin penetrating connective tissue septa divide the corpuscles into several incompletely delimated lobes. The third type septa are better developed and more prominent then the previous type. The union of ramifying connective tissue causes same groups of cells to become separated from the rest, forming smaller lobular from the completely delimated lobes. In the fourth type corpuscle is composed of aggregetes of small lobes. Each of which consist of a number of complete and incomplete lobes. The penetration of connective tissue septa into single corpuscles makes the corpuscles bilobed, trilobed or even multilobed condition.Two principle types of secretory cells are observed. Type-l (PAS+ve) and Type-2 (PAS-ve). Type-1 cells are round with clear vesicular nucleus and stainable material in the cytoplasm Type-II cells are slender, irregular cell bodies that may contain cytoplasmic processes extending between the type-I cells, A third type-III cells was also noticed.

Histochemical Studies

Histochemical studies for the demonstration of lipids, proteins and dichromate reaction for the localization of any adrenaline and nor-adrenaline containing cell was carried out. The presence of lipids was detected by Sudan Black-B staining method. The corpuscle was fixed in calcium formol and stained with Sudan Black-B. In presence of Sudan black color in some portion of the Stannius corpuscles indicate the presence of small amount of lipid.The presence of proteins was observed by Mercury-bromophenol-blue staining method. The corpuscles showed the blue color, which indicated the presence of proteins. The cells of corpuscles are well-equipped for protein synthesis and secretion. The positive reaction indicates presence of glycoproteins which may be stanniocalin as reported in other fishes.

To detect the presence of adrenaline and nor-adrenaline containing cells, the corpuscles of Stannius was fixed in potassium dichromate solution. The presence of some dark brown and light yellow pigments shows the presence of pigment cells. The positive reaction to dichromate was observed in between the CS cells indicating innervation.

Ultrastructural Studies

The ultrastructural observation of corpuscles of Stannius in the fish, *Notopterus notopterus* shows basically three types of cells. The identification of three different cells is based on the nuclear structure, morphology of secretary granules and other cytoplasmic components (Fig.4.5). The semithin sections prepared for ultrastructural studies shows the cellular organization (Fig.4.6). At the ultra structural level, cells of three types are located in the CS of *Notopterus notopterus.* These secretary cells as type-1 are arranged in follicles (Fig.4.7). They are characterized by oval or round nuclei with uniform chromatin material. and numerous large membrane bound electron dense secretary granules scattered all over the cytoplasm. Abundant ribosomal endoplasmic reticulum (RER), arranged as lamellar arrays in the vicinity and nucleus as well as in the rest of the cytoplasm are also visible. A few elongated filamentous mitochondria with lamellar or tubular cristae are present in the cytoplasm. The Golgi areas are well developed with large vacuoles containing electron dense materials and some are completely empty. The type-II cells are characterized by electron dense cytoplasm, few membrane bound secretary granules, the nucleus has chromatin patches and these cells have cytoplasmic extensions amongst the type-I cells, the type-III cells are smaller in size and lesser in number with uniform cytoplasm having less secretary granules.These cells have broader nucleus with chromatin material. Amongst the three types of cells clearly identified, the type-I cells are prominent and found to be the major cell type probably secreting the hormone of the corpuscles of Stannius. Amidst the gland cells, a mast cell packed with membrane bound electron dense bodies of variable size and a large irregular shaped nucleus is also conspicuous.

Identification of the Hypocalcemic Factor

Figure shows the densitometric scans of coomassie brilliant blue stained products after SDS-PAGE under reducing conditions present in a crude tissue homogenate of CS of the fish, *Notopterus notopterus.* A product with an apparent molecular weight of approximately 41 kDa Protein analysis from the tissue homogenate of 100 mg dry weight of CS with a protein content of 6 mg.

SDS-PAGE: (SDS-PAGE Analysis of Partially Purified STC from the fish *Notopterus notopterus*

Lane-1 is molecular weight morker;phosphorylase (97 kDa); bovine serum albumin(66kDa);ovalbumin(43kDa);carbonic anhydrase(29kDa);lane-2 crude, lane-3 ammonium sulphate, lane-4 after dialysis, lane-5 G-100.

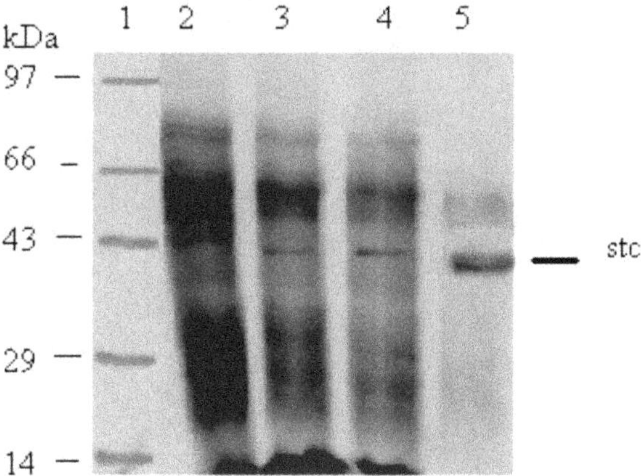

Fig.4.5: SDS-PAGE analysis of partly purified steniocalcin in the corpuscles of Stannius extract of The fish, *Notopterus notopterus.*

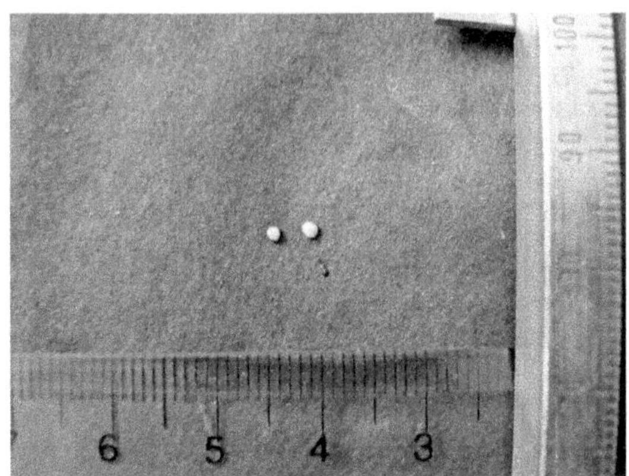

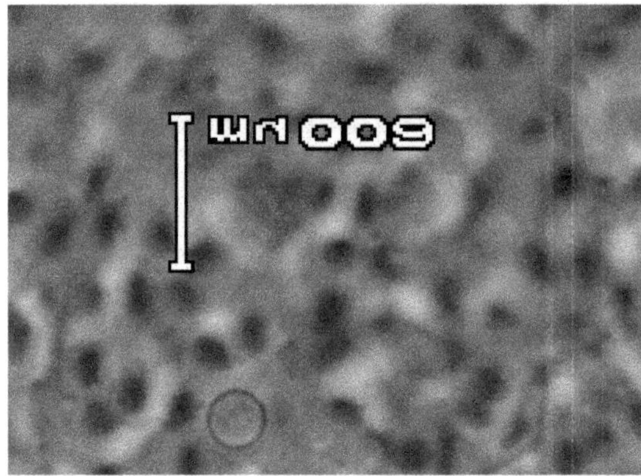

Fig.4.6: Photograph below showing the whitish corpuscles of Stannius of *Notopterus notopterus* and section of corpuscles of Stannius composed of columnar cells, closely packed in groups.H & E. ×1200.

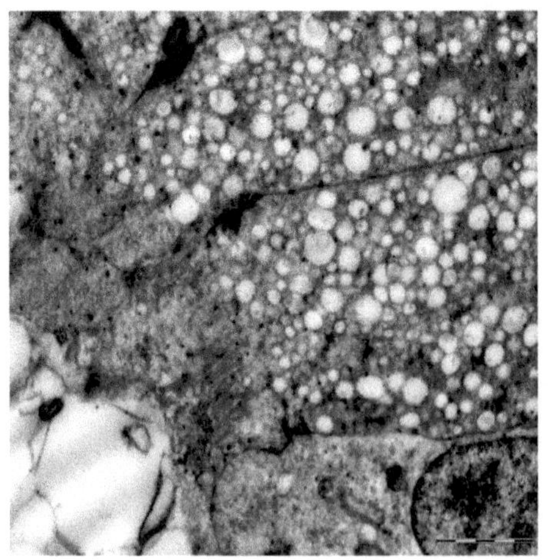

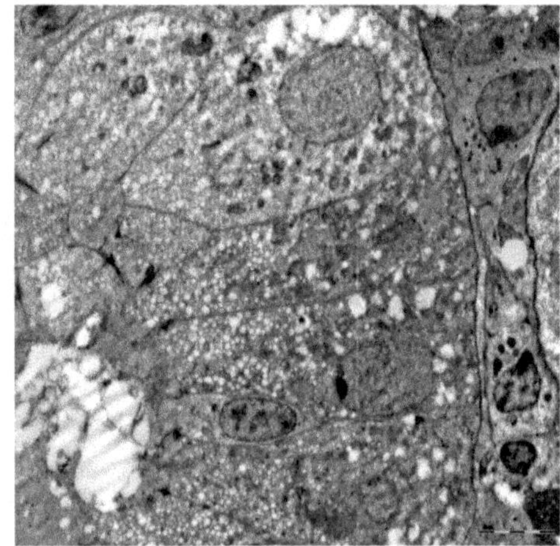

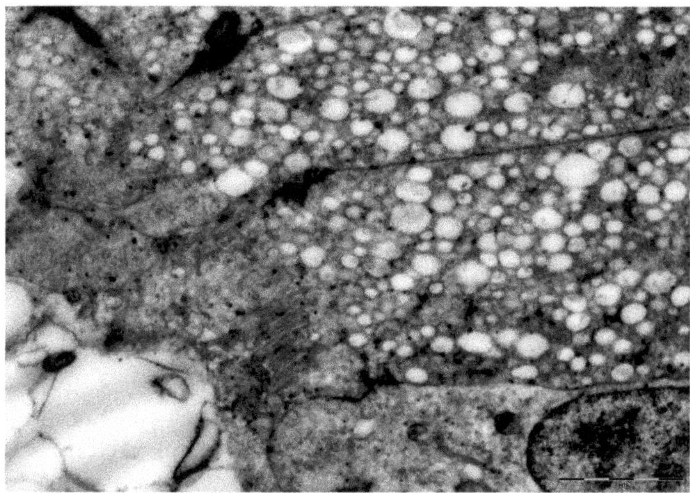

Fig.4.7: Ultrastructure of corpuscles of Stannius below in the fish, *Notopterus notopterus* showing highly vacuolated type-1 cells × 9300., Section of the corpuscles of Stannius showing ultra structural details of the type-I cells × 2900.

Hormone of the Corpuscles of Stannius

The corpuscles of Stannius present in the holosteans and teleosteans are specific endocrine structures. Their functional significance remains unsettled for a long time. These glands arise embryologicaly as evaginations of the wall of the pronephric ducts. They are not homologous with adreno-cortical tissues which are formed from the coelomic mesoderm. In teleost fishes, the corpuscles of Stannius can be found either as a single pair of symmetrical structure at the dorso-posterior end of the kidney and they occur in variable number in different species of fishes. However, in most cases the number is limited 2-6. In number of telostean fishes posses a single pair of corpuscles. In the fish *Oncorhynchus gorbuseha* a pair of corpuscles of Stannius has been reported. Yet in another two species of *Oncorhynchus; Oncostymchus Tshawytstha* and *Onclorhymchus* Kisutch, have observed 5-6 corpuscles. Compared to holosteans the number of corpuscles in teleosts is usually lower, although up to fourteen have been reported for salmonids. In gold fish or stickle back (*Gasterosteus aculatus*) only a single pair can be found at the dorso-posterior end of the kidney. The number of peripheral corpuscles has been shown to vary from 4-6 in *Salmo gairdnerii* 4-5 in *Oncorhynchus* Kisutch, 2, 3, 4, in *Atherinopsis californiensis,* three in *Sepastobes aurinsbica.* The number of CS may vary from 4-10 in *Salnio solar* and 6-8 in *Salmo trutta.* It has been also reported in many cases (examples catfishes). The organs become reduced in the adult to single corpuscles. In the earlier reports published by Belsara (1973) only one corpuscles of Stannius is present in the fish *Notopterus notopterus.* In our study on the same species a pair a corpuscles of Stannius are present and located in the anterior region of the posterior kidney. In other Indian fishes such as *Heteropreustes fossilis* as many as 4 corpuscles are frequently observed, although one may be present in the same species, it has also pointed out in catfish *viz. Heteropneustes fossillis, Clarias batrachus* and *Mystus vittatus*, the number of corpuscles of Stannius is more than one. After a comparative study of corpuscles of 47 species (Bauchot, 1953), concluded that a phylogenetic significance can be attributed to the number and location of the CS. In more primitive fishes the number to be high and glands are often located in the middle of the meso-nephric kidneys. However, many exceptions to these rules have been reported in a rather primitive species like *Notopterus notopterus* (order; *Clupiformecs*) only one corpuscle of Stannius is present located in the posterior position (Belsara, 1973). Where as in the more advanced representatives of cypriniformes large number may occur, often in more anterior position, upto 10 may be present in *Clarias batrachus*, upto 8 in *Heteropheustes fossilis* and between 2 and 4 in *Mystus vittatus.*

Moreover, within one species individual differences may occur that are related to age or sex. In *Esox lucius* and *Colisa lalia* the corpuscles of Stannius may be reduced in number from 5-2 during growth of the fish. Where as, in the *Chilean ching* fish *Sicyasis sanguineus*, the males have two corpuscles while females usually have three corpuscles. In the present study, the fish *Notopterus notopterus*, a single pair of corpuscles of Stannius were found in both sexes embedded in the anterior portion of the posterior kidney. The variation in the number of corpuscles of Stannius in the same species reported by earlier worker seems to be an embryological speciality rather than a matter of evolutionary or taxonomic significance as suggested by Belsare (1973). There are also some reports that in the same species such as *Heteropneustes fossilis* as many as four corpuscles are frequently observed although one may be present in some.

The CS were long considered to be the homolog of the adrenal cortex of higher vertebrates because of their location and their superficial resemblance to the *Elasmobranch interrenal.* However, it was later shown that the embryological origin of corpuscles of Stannius was completely different from that of the inter-renals. In teleosts it was reported that CS are derived from the pronephric and mesonephric ducts. In several salmonids and in more advanced species like *P. mattatus*, in catfish and *Lolisa lalia*. The corpuscles of Stannius are formed from evaginations of mesonephric ducts and this embryological origin makes the CS unique among endocrine gland of vertebrates. The location of corpuscles of Stannius in the teleostean kidney presents several variations and this variation in location suggested to be related to the taxonomic position of the fish species. In *Notopterus notopterus* in the present study the CS occur at the anterior end on either side of the posterior position of the kidney but in majority of the fishes they are located at the posterior third of the meso nephron. In holostean the CS occupy the area near the anterior end of the pronephric duct, in isospondyles they are situated in the mid region, while all the majority of the fishes they have moved back to the posterior third of the mesonephros, has pointed out that corpuscles of Stannius move progressively backwards during the evolution as a result of actual shortening of body cavity rather than a migration of corpuscles of Stannius. Similar observation has been also made in *Heteropneustes fossilis* by Subhedar and Rao (1976), and suggested that it appears that this species (*H. fossilis*) occupies a higher evolutionary status than that of isospondyles. However, in the *Notopterus notepterus* the CS are located at the anterior end of posterior kidney, it is possible that this fish may at the lower status than that of cat fishes.

In *Notopterus notopterus* in the present study, the corpuscles of Stannius are oval in structure, the cells of the glands are usually arranged in strands, lobes or lobules that are separated by septa of connective tissue continuous with the outer fibrous capsule.The septa contains all vascular and nervous elements. After the survey of the organisation of the corpuscles of Stannius in 29 species which has been divided into four categories on the basis the arrangement of the gland cells (varying from major lobes to small aggregate of gland cells) and the extensiveness of the connective tissue septa. Later observations have shown that in some species the arrangement of the gland cell may differ between the CS of individual fish and more importantly that in many fish the arrangement of the gland cells and the extensiveness of the septa depend on the state of activity of the glands. The cytoarchitecture of the CS of the catfish *Heteropneustes fossilis* has been investigated by Subhedar and Rao (1976) based on the arrangement of septa and cells and reported that at least 4 principle architectural patterns can be distinguished. In the first type, the cells appear either as cords or as follicles due to complete formation of thin septa in the second category also. The penetrating septa are thin but divide the CS into incompletely delimited lobes. In the third type, the septa are thick and their branches united, some times resulting in the formation of complete lobes. In the fourth type, each CS is formed of aggregates of lobes each of which consists of a number of complete or incomplete lobules. Such similar observation has been made in other fishes. On critical observation in the light of evaluation made by others all the four types of CS has been noticed in the CS of *Notopterus notopterus* in the present study.

The vascularisation of the corpuscles of Stannius in the *Notopterus notopterus* is rich this study is based on the observation made through histological histochemical and electron microscopic studies. It is also reported that vascularisation of the CS is usually rich and it may become even more elaborate when the secretory activity of the corpuscles of Stannius is increased. This leads to an extensions of capillary network in the connective tissue septa and a subdivision of The glandular lobes, which results in a better access of the gland cells to the blood circulation. The blood capillaries in the septa are usually of the frustrated type.Data on the vascular supply to the corpuscles of Stannius are limited in a mullet (*Mugil cephalus*) the blood supply of the corpuscles of Stannius is associated with the renal arteries and the cordinal veins, in the corpuscles of Stannius of stickleback (*G. aculeatus*) the vascular elements from a portal system, similar to the blood supply of the kidney. The corpuscles of Stannius receive their blood from the dorsal caudal veins and from segmental veins from the dorsal musculature. They branch into an extensive network that is drained by veins opening into the renal blood circulation. These studies indicates that the product released by the corpuscles of Stannius first pass through the kidneys before they enter the general blood circulation. The corpuscles of Stannius also have a rich nervous supply, as was already noted by Huot, (1998).

In a light microscope study of eight teleost species, it was demonstrated rich autonomic innervations of corpuscles of Stannius, but they were unable to show whether it was sympathetic or parasympathetic. In all species examined, nerves and ganglia were found in close proximity to the corpuscles. Ganglia like accumulation of neurons usually in close proximity to the corpuscles but occasionally within the corpuscles capsule.From this ganglion, nerve fibers penetrate the bodies along the blood vessels and branch into small fibers that run into the septa in close association with the capillary network. In many species the septa also contain single neuron. The nerve fibers are always confined to the septa. They do not seem to penetrate between the glandular cells and synaptic nerve endings on gland cells have never been reported.

The lack of evidence for direct nervous innervation of the gland cells together with close association invariably found between the nervous and vascular supply of the corpuscles of Stannius have led most authors to conclude that the nervous supply of the CS is from a primarily connected with the control of the blood flow through the bodies any effect on the glands may be indirect. In the most detailed study of the nervous supply of the CS available so far, Unsicker *et al.,* (1977) came to the same conclusion for rainbow trout (*Salmo irideus*). With flourimetric technique they showed that the corpuscles of Stannius contain considerable amount of catecholamines and 5-Hydroxytryptamine they also shown that the nerve fibers running in the septa are adrenargic, although the fibers also might contain some 5-hydroxy tryptamine. The neurons dispersed in the connective tissue appeared to produce nor-adrenaline. These cells were sympathetically innervated by cholinergic nerves.

The ultrastructural observation of CS cells in the *Notopterus notopterus* shows three types of cells. The first type cells are type-I cell are large in number having electron transient cytoplasm with vacuolation, large secretary granules and abundant mitochondria and ribosomal endoplasmic reticulum, the second type of cell known as type-II are less in number characterized by electron dense cytoplasm, few membrane bound secretary granules, these cells have cytoplasmic extensions amongst the type-I cells. The third type of cells also have been observed laying small in size having broader nucleus, well marked nuclear membrane and has uniform cytoplasm with very less secretary granules.

The most conspicuous feature of these cells particularly type-I cells is the presence of a large number of vacuoles of irregular shape which gives exhausted appearance. The secretary granules are large in type-I cells where as, they are smaller in type-II cells. The Golgi areas are seen, the mitochondria are large and more in number in every cell particularly in the type-I cells the nuclei of all the three types of cells are clear with prominent nucleolus. The chromatin particles are seen in the type-II and type-III cells, which are not seen in the type-I cells. Hence, based on the structure of nucleus and cytoplasm, the three types of cells can be easily distinguished.

The electron microscopic observations have been carried out in CS cells of other fishes and it has been reported that the freshwater fish species possess two types of cells (type-I and II) in their corpuscles which the marine fishes have just one type. However, exceptions have been reported with regard to types of cells in the CS of teleostean fishes in the fish, *Notopterus notopterus,* and three types of cells are clearly visible based on the cytoplasmic state, nuclear structures and secretary granules. The prominent cell type is type-I cells which have large secretary granules with abundant mitochondria and extensive vacuolations suggesting that major hormone of the CS may be secreated by these cells in comparison to other two types of cells (type-II and III). The CS of teleosts residing entirely in freshwater or euryhaline species, spending part of their life cycle in freshwater have been shown to contain heterogeneous populations of gland cells, unlike the marine forms where the CS contain cells of only one type (Wendelaar Bonga and Grevens 1975), *Onchorhynchus mykiss* (Krishnamurthy and Bern 1969; Meats *et al.,* 1978) *Fundulus heteroclitus* and *Carassius auratus* (Wandelaar Bonga., 1980), *Oreochromis mossambicus.*

The mast cells of *Notopterus notopterus* are located adjacent to the Stannius cells under EM observation and are closely associated with the blood capillaries. The cytoplasm of the mast cell is packed with large number of membrane bound electron dense bodies of variable sizes. The nucleus of mast cell is large and irregular in shape. The presence of mast cells in corpuscles of Stannius has been also noticed in the fish, *Heteropneustes fossilis* and the authors have suggested that the mast cells are phenotypically and functionally varsatile effector cells. It is also known that mast cells produce and release a diverse array of mediators like 5-hydroxytryptamine, heparin, serotonin, histamine, lipeid mediators etc. The presence of mast cells at strategically important locations i.e., near the blood capillary and secretary cells is suggestive of a functionally important role. The above authors have further suggested that mast cell is acting as an intraglandular fine tuner for the secretion of stannicalcin in paracrine fashion.

The linkage of mast cell in the secretary process may be calcium of the blood stimulates mast cell to release mediator serotonin or other substances which acts directly on the Stannius cells or alternatively causes vasodilation resulting in increased flow of blood, thus carrying the secretary product at an enhanced rate and thus location of mast cell gives it a unique opportunity to influence both release and transport of staniocalcine. In the present study groups of mast cells noticed near the blood capillaries. So, cells of corpuscles of Stannius in the fish, *Notopterus notopterus* may be having similar kind of activity as reported by Feroz Ahmed *et al.,* (2002).

The study on the SDS-PAGE application of CS tissue homogenate indicates that the product of CS of *N. notopterus* is a protein having a molecular weight of 41 kDa which might be a hypocalcemic hormone reported in other fishes. Because of its high molecular weight, this product from CS of the fish *N. notopterus* can be called as hypocalcin.

As this name (hypocalcin) was proposed for a CS hypocalcemic principle with a molecular weight about 10 kDa by Pang *et al.,* (1974,), whereas the name teleocalcin was first given to a hypocalcimic principle with a molecular weight of 3 kDa isolated from CS of Salmon.Hypocalcin is present in relatively large amount in the CS of the six freshwater species such as European eel, Tilapia, gold fish and carp after SDS-PAGE under reducing conditions found that the product with an apparent molecular weight of approximately 54 kDa and concluded that after SDS-PAGE under reducing conditions the product from CS of the fishes appears as a dimer identified as a 41 kDa band.

The present observation on the molecular weight of 41 kDa for the product (hypocalcemic principle) of *N. notopterus* were in agreement with the reports of Lafeber *et al.,* (1988) in six species of fish studied showing that isolation of hypocalcin yields a band of 41 kDa after SDS-PAGE under non-reducing conditions. There are similar reports on other fishes. Hence, it is indicated that the CS of the freshwater fish *N. notopterus* secretes a product having molecular weight of 41 kDa as that of other fishes and this product of CS can be called as hypocalcin hormone.

Some Experimental Studies on the Corpuscles of Stannius (CS) and Serum Calcium Level in the Freshwater Fish, Notopterus notopterus

The Corpuscles of Stannius (CS) have been most consistently implicated in the control of plasma calcium metabolism. Stanniocalcin (STC), the hormone secreted by the CS, lowers plasma calcium levels by reducing gill calcium uptake. Histological examination of the CS has revealed that the glands are consistently more active in seawater-adapted fish than in freshwater adapted fish. In addition, CS of fish kept in calcium deficient sea water appears to be inactive suggesting that glandular activity is directly related to environmental calcium concentrations. The corpuscles of Stannius were recognized as endocrine glands of a specific nature of the two main functions attributed to the CS are control of calcium metabolism and cardiovascular action. The CS is small ovoid endocrine glands unique to teleostean and holostean fishes, embryologically the CS may arise from the pronephric, mesonephric or opisthonephric ducts. The corpuscles of Stannius (CS) are enigmatic glands which vary in number and are usually embedded in the kidneys. The role of CS in fish has not been still established. However, there are some reports that CS secretes a glycoprotein hormone which regulates calcium homeostasis. Hence, in the present investigation effect of CS extract on calcium regulation has been studied in the freshwater fish *Notopterus notopterus.*

The fish, *Notopterus notopterus,* weighing 90-100 ± 10 gms were collected and were selected and divided in to three equal groups. All were sexually mature, between two to three years of age and appeared in healthy and vigorous condition. The first (I) group fishes maintained in natural water as control. Second (II) group fish were maintained in calcium rich water (0.1% $CaCl_2$) solution both group-I and II injected with (0.5 ml / 100 gm bw) 0.6 mM NaCl solution and the third group fishes were maintained in calcium rich water (0.1% $CaCl_2$) and 0.5 ml CS extract injected.

The CS tissue was collected from mature fish and homogenized in a morten with appropriate volume of saline solution. Homogenate was centrifuged at 3,000 rpm for 10 min and the supernatant was filtered and used. All the experimental and control fishes were netted from the water and blood was taken via a syringe inserted in to the intraperitoneal. The blood sampling was completed within a period of 2 minutes and by using the modified method by Clark and Collip was employed for estimating the calcium in the serum, the result obtained being mg. Ca per 100 ml serum. The fish were killed by decapitation. In each group of three or more animals the body cavity was opened and the kidney removed. A standardized region of the musculature [Full flank dorsal to the lateral line] was removed from the fish and all the tissues fixed in the Bouin's fluid. About an hour later the kidney, containing the corpuscles of Stannius, were dissected out and returned to fresh fluid. This procedure insured rapid fixation of the rather labile granules of the glandular tissue followed by dehydration in graded alcohols and embedded in paraffin blocks using standard techniques, five to six micron thick sections were cut, dried and dehydrated for Harri's hematoxylin and eosin stain. Microphotographs were taken using Olympus DP-12, Olympus BX 51, and Model-ULH 100 HG, Olympus Optical Co. Ltd. Made in Japan.

The serum calcium level estimated in the blood for different experimental groups are presented. The changes in the histological structure of the CS under different experimental conditions have been also studied. The functional role of corpuscles of Stannius in fish remained unsettled for long time. However their role in osmo-regulatory action has been accepted now by the endocrinologists after experimental results from time to time. Now it is acknowledged that Stannius of corpuscles produce hormone(s) which restrained the development of post stanniectomy hypocalcemia and correct the raised serum calcium which develop after stenniectomy the present study is carried out by exposing the fish *Notopterus notopterus* to calcium rich water and injected with CS extract. The serum calcium level has been determined. With recent advancement made in understanding the role of CS in calcium homeostasis shows that the secretion of CS is known as Stenniocalcin - a glycolprotein has a biological role in the inhibition of gill calcium transport and reduce calcium absorption.

Secretion of CS depends on calcium as stimulating factor this study is based on the intraperitoneal injection of calcium chloride resulting in CS cell degranulation, synthesis and secretion of the hormone in fish *Notopterus notopterus* after exposing calcium rich water.

The functional role of corpuscles of Stannius in fish remained unsettled for long time. However, their role in osmo-regulatory action has been accepted now by the endocrinologists after experimental results from time to time. Now it is acknowledged that Stannius corpuscles produce hormone(s) which restained the development of post-stanniectomy hypercalcemia and correct the raised serum calcium which develop after stennioctom. In the present study, after exposing the freshwater fish *Notopterus notopterus* to calcium rich water has shown increase in serum calcium, when these fishes injected with CS extract reduced serum calcium indicating that CS extract containing the active compound causes reduction in the serum calcium as reported in other fishes. With recent advancement made in understanding the role of CS in calcium homeostasis shows that the secretion of CS is known as Stenniocalcin - a glycolprotein has a biological role in the inhibition of gill calcium transport and reduces calcium absorption. Secretion of CS depends on calcium as stimulating factor.

This study is based on the intraperitoneal injection of calcium chloride resulting in CS cell degranulation, synthesis and secretion of the hormone in fish *Notopterus notopterus* after exposing calcium rich water up to 15 days the serum calcium level increased and was higher, whereas after 15th day up to 30th day the serum calcium was lower, suggesting that in the freshwater fish *Notopterus notopterus* also calcium regulating factor is being secreting from CS causing inhibition in the absorption of calcium leading to reduction in the serum calcium level. In the CS extract injected fishes also the serum calcium level reduces after 15th day bringing to normal level as that of control fish such similar observation were made in other fishes. The microscopic structures of CS cells have been studied in other type of fishes. After exposing calcium rich water, their results clearly state that calcium environment leads to STC secretion from CS through their observation on cellular details especially enlarged rough endoplasmic reticulum and enhanced exocytosis. In the present study also the CS cells of *Notopterus notopterus* are active after exposing to calcium rich water and CS extract injected indicating similar kind of hormone (STC) may be secreted from CS cells regulating serum calcium level.

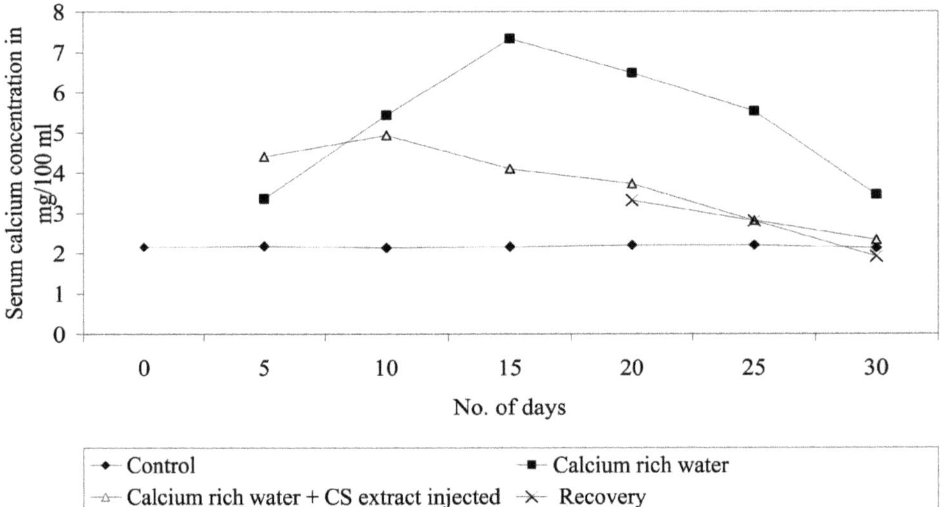

Fig.4.8: Serum calcium level at different experimental conditions in the freshwater fish, *N. notopterus*

Chapter-5

Structure of the Pancreas and its Hormones

The pancreas or pancreatic tissue is large and important gland in the body of an animal and it lies in the abdomen with its head resting in the concavity of the abdomen. This gland is both exocrine and endocrine in function. The morphological structure of the pancreas variable among teleost fishes.However, many phylogenetically "higher" teleost fish species have isolated masses of one or two islets which are called Brockman bodies. The structure of the endocrine pancreas is unique in its tendency to form either a large accumulation of islet tissue (principle islets) or a similar accumulation surrounded by a rim of exocrine tissue. This morphological peculiarity of fish pancreas constitutes a valuable model for investigation of islet hormones and their interactions. Recent reviews reflect an increased interest in the evolutionary and developmental aspects of the endocrine pancreas in vertebrates. Insulin is an important hormone secreted by the beta cells of endocrine pancreas. It helps to regulate the blood glucose in the body. There are reports available that concentration of glucose in the blood varies with season and glucose regulation is dependent on the beta cells volume in all major groups of vertebrates. In this chapter the location, morphology, histological structure of pancreas and variation of insulin hormone during different season of a year in the fresh water fish *, Notopterus notopterus* was studied and presented.

Haematoxylin and Eosin stained serial paraffin embedded sections were used to identify the location and distribution of the pancreas of the fish, *Notopterus notopterus.*The general morphology and the distribution of the pancreas was confirmed most of the earlier findings by the present study.

The pancreatic gland tissue in the fish, *Notopterus notopterus* lies in the abdomen with head resting on duodenum and from the duodenal base, the compact pancreas split into two lobes and runs along the dorsal and ventral portion of the intestinal outgrowth, which are characterised by two parts, firstly exocrine acini and their ducts and secondly endocrine islet tissues. Endocrine islet tissue was always found closely associated with endocrine pancreatic tissue. In the exocrine pancreas darkly stained irregularly arranged acinar cells and lightly stained islets are observed.

The islets of various sizes were widespread throughout most of the pancreas region. Inside the islets, large cells at the periphery and smaller cells in the middle of the islets were noticed. Each islet was enveloped by a delicate connective tissue capsule, which separated it from the surrounding layers of exocrine tissue.The pancreatic lobes of *Notopterus notopterus* have packed acinar cells arranged in most irregular way. Abundantly seen acinar cells separated by thin film of reticular tissue surrounded by scattered islets of Langerhans. The islets contain larger alpha cells at the periphery and smaller beta cells in the middle region.

Based on microscopic observation it has been suggested that the pancreas of the fish, Notopterus notopterus is of tetrapodan type and is located in the concavity of the duodenum with its head resting and the body of pancreas is split into dorsal and ventral portions. From the literature study it was reported that there is fundamental difference in pancreas structure and histophysiology between cyclostomes and gnathostomes. Pancreas morphology and islet structure of all higher vertebrates can be derived from primitive gnathostome type of the pancreas as exemplified by some elasmobranchs.

The histological structure of pancreas contains the various endocrine cells that have been recognised in other modern fishes,the teleosts. These different types of cells are also secrete hormones such as insulin,glucagon, somatostatin and pancreatic polypeptide which needs further verification as far as fish tissue is concerned. The histological study of the pancreas clearly showed the presence of islets of langerhans, β cells are responsible for the secretion of insulin hormone, which are present in islets of langerhans of endocrine pancreas.

The insulin hormone estimation was performed for the fish, *Notopterus notopterus* by applying coat- A –count insulin 125 I radioimmunoassay (RIA) method during four reproductive phases such as preparatory, prespawning, spawning and post spawning phases. The serum insulin level determined was found to be between 3.916 to 5.466 mlu/ml. And the blood glucose level ranges from 26.50 to 59.50 mg/dl. Hyperglycemia is a usual response to insulin in teleosts and a deficiency in insulin permits a rise in plasma glucose and other metabolites as reported. In the fish, Notopterus notopterus both insulin and glucose level was high during spawning phase compared to other reproductive phases.The insulin level increases from preparatory phase to spawning phase may indicate the requirement of hormone and glucose during breeding period (Spawning phase) for increased metabolic activity as the fish has to perform reproductive effort of gametogenesis and release of gamates. The amount of insulin and glucose level during different reproductive phases confirmed that there is variation of hormone level during different reproductive phases of the fresh water fish *Notopterus notopterus.* The results suggests that the fish, *Notpterus notopterus* has distinct islets of langerhans in the endocrine pancreatic tissue as that of higher vertebrates and is capable of synthesising insulin hormone regulating blood glucose level. The location , morphological , histological structure of pancreas and islets of Langerhans along with insulin/glucose level determined is presented in the fresh water fish, *Notopterus notopterus.*

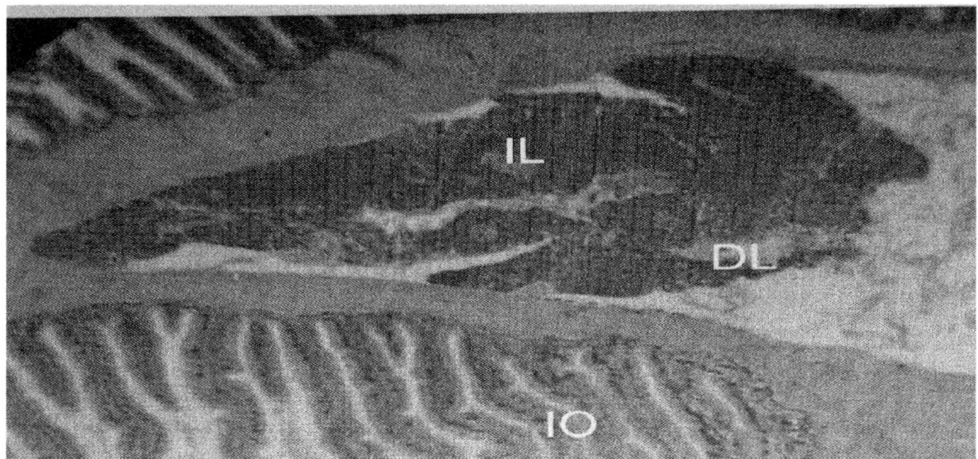

Fig.5.1: Two lobes of pancreas of the fish, *Notopterus notopterus.*

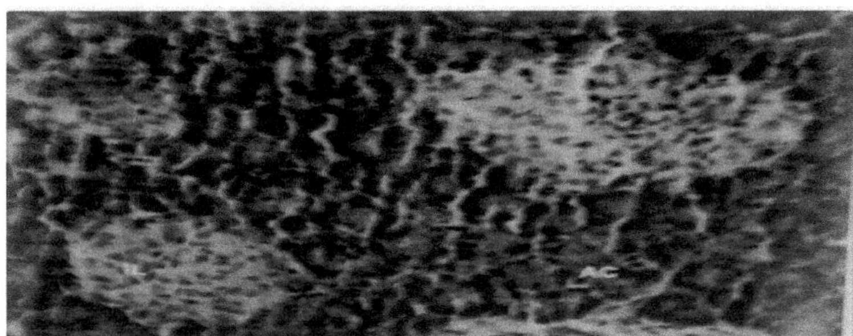

Fig.5.2: Histology of pancreas showing islets of Langerhans in the fish, *Notopterus notopterus.*

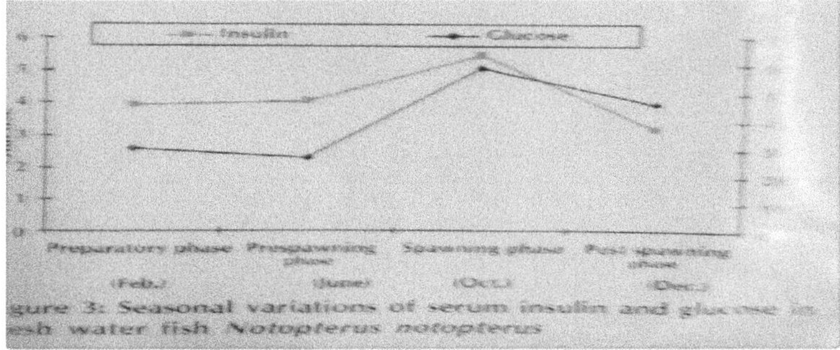

Fig.5.3: Seasonal variations of serum insulin and glucose level in the fish, *Notopterus notopterus.*

Chapter- 6

Gonads and Gonadal Hormones

Structure of Testis and Testicular Cycle

The histological preparation of the testis in *N. notopterus* consists of large number of lobules/tubules in varied shapes and sizes. The testicular germ cell nests are separated by fine connective tissue. The lobules cysts undergoing different stages of development are clearly seen. The lobulated testis internally has somatic cells in addition to germ cells. These somatic cells are Sertoli cells and interstitial Leydig cells. The cysts of the lobules contain different spermatogenic stages in the fish *N. notopterus* six spermatogenic stages were identified. They are primary and secondary spermatogonia, primary and secondary spermatocytes, spermatids and spermatozoa. The primary spermatogonia are larger in size with a distinct round to oval vesicular prominent volumnous nucleus, the cytoplasm of these cells is homogenous. The secondary spermatogonia are smaller than the primary spermatogonia each containing oval to round nucleus. The primary spermatocytes are smaller than spermatogonial cells having round small nucleus. The nuclei of these cells are under different stage of divisions. The chromosomal configuration appears to be banquet type. The secondary spermatocytes are still smaller cells compared to primary spermatocytes with little cytoplasm.The nucleus stains intensely with hematoxylin. The spermatids are still smaller cells having distinctly stained nucleus. The cytopalsm is acidic. These cells form separate group inside the lobules. Spermatozoa are the smallest cells of the spermatogenic cells and are produced from spermatids. They are concentrated in the lobular lumen. These cells have a well differentiated head and long tail. The active motility of these cells was observed in the milt extracted during spawning phase.

The Seasonal Testicular Cycle

The testicular cycle studied during one year period indicate that it passes four reproductive phases similar to ovarian cycle. This study is based on the computation of TSI, histology of the testis and marphometric studies based on the lobular diameter. The four phases in which the testis underwent changes are categorised in to preparatory phase, prespawning phase, spawning phase and post spawning phase.

In the preparatory phase the histological section of the testis consists of lobules having primary and secondary spermatogonial cells and actively dividing primary and secondary spermatocytes. Some sperms are also seen. In the prespawning phase the gametogenic activity increases resulting in the increase in the concentration of spermatozoa in the lobular lumen. The TSI also increases. The primary spermatogonia are decreased, the primary and secondary spermatocytes were more numerous during initial two months of this phase and during later months they were less as they have converted into spermatozoa. In the advanced stages of testis during prespawning, the testicular inter tubular connective tissue septa shows thickening and disorganised. During July and August the seminiferous lobules broke down and as result the spermatozoa form a mass showing sperm reservoir.

In the spawning phase, the TSI decreases, the primary and secondary spermatogonia increased for nearly 1% in August to 40% in September. The lobular lumen either packed with spermatozoa or in some it is empty. The post spawning phase testis is characterised by the presence of spermatogonia and spermatocytes which are not in the process of divisions.

The observation made based on the TSI, histology of the testis strongly suggests that the testis undergoes annual rhythm of spermatogenesis. The data indicate that the spermatogenic activity was highest during pre-spawning phase (April-July) and spawning occurs during August-September. The mature stage of the testis is seen during prespawning and spawning phase while immature testis was found during post spawning phase. In the preparatory phase the testis is preparing for spermatogenic activity after it has undergone a resting phase during post-spawning phase. Gonadosomatic index:The gonadosomatic index determined in both male and female fish presented in the Table, showing that it increases from preparatory phase to prespawning phase and further decreases after spawning in the post spawning phase. This indicates that the gonads undergo increase is weight during prespawning phase and depletion after spawning in the post spawning phase.

Testicular Steroid Synthesising Cellular Sites

They can be identified in histological preparation during prespawning and spawning phases. The Sertoli and interstitial Leydig cells exhibit different characteristic, the sertoli cells are large and are inside the lobuler lumen closely attached to cysts undergoing spermatogenic divisions, whereas the interstitial Leydig cells are smaller and are arranged in groups in between the lobules.-

Testicular Steroid Hormones

Testosterone: In male *N. notopterus* the plasma concentration of testosterone (mg/ml) for four reproductive phases was studied. Male fish plasma testosterone levels peaked during spawning phase and dropped to a lower level in post spawning phase. The level of the hormone during preparatory and prespawning phases increased and remained constant during these two phases. Maximum increase is seen only during spawning phase. The plasma level of testosterone in female fish

N. notopterus though the concentration of the hormone is less in females than males, peak level found to be during spawning phase and remained almost constant in other phases with minimum level during post spawning phase. The higher level of the hormone testosterone in male fish during three phases suggests its role in spermatogenesis.

Structure of the Ovary and Ovarian Cycle

The ovary of *N. notopterus* is covered by a thin layer of peritorium. The peritorium internally attached to tunica albugenia which projects as ovigerous lamellae into the ovocoel. The ovary consisting of different stages of oocytes embedded inside the loosely connective tissue. The different oocytes of the ovary have been classified into nine stages. They are (i) oogonium (ii) chromatin nucleolus (iii) early perinucleolus (iv) late perinucleolus (v) cortical alveoli (vi) primary yolk globule (vii) secondary yolk globule (viii) tertiary yolk globule and (ix) Migratory nucleus. The oocytes beloging to oogonium, chromatin nucleolus and early perinucleolus stages were present in the ovary throughout the ovarian cycle whereas other stages of oocytes are present during prespawning and spawning phases. The different stages of oocytes exhibit characteristic features in their cellular components such as:

i. Oogonium: These are embedded in the ovigenous folds and can be identified by their small, round to oval in shape.The size of this stage is 5 to 6 µm in diameter and found in groups present in between growing oocytes. They have single nucleolus and the cytoplasm is weakly eosinophilic.

ii. Chromatin Nucleolus: These stage oocytes are abundantly seen during preparatory phase. They are larger than oogonial cells having round to oval in shape the mean diameter is around 20.05 µm. The cytoplasm is homogenous with a central nucleous.

iii. Early Perinucleolus: The oocytes are larger than the chromatin nucleolus oocytes; they can be differentiated based on the small size and a prominent nucleus having conspicuous nucleolus. The diameter of this stage oocyte is around 24.30 µm. The cytoplasm is homogenous with a nucleus having scattered nucleoli also arranged attached to nuclear membrane.

iv. Late Perinucleolus: This stage of the oocytes are characterised by the presence of yolk nucleus in the homogenous cytoplasm stained with hematoxylin and eosin. The follicular layer has differentiated and noticed around the oocyte. The nucleoli increased in number and arranged at the periphery of the nucleus. The diameter of this stage oocyte is around 37.45 µm.

v. Cortical Alveoli Stage: The diameter of this stage oocyte is increased in comparison to the earlier stage oocyte and found to be 40-48 µm. These oocytes are characterised by the presence of alveoli distributed in the cytoplasm, these alveoli are present in the periphery of the ooplasm and in the later stage the number and size of the alveoli increases. The nuclear contents are homogenous and finally granular.

vi. Primary Yolk Globule Stage: This stage oocyte is characterised by the presence of large number of alveoli forming a network appearance. The average diameter of this stage oocyte is 48.92 µm. This stage may indicate the endow vitellogenic activity.

vii. Secondary Yolk Globule Stage: This stage oocyte diameter is increased to 49.07 µm. The cortical alveoli were found closely packed inside the cytoplasm and also at the periphery of ooplasm. They form ring like band around the remaining portion of the ooplasm. The ooplasm is completely filled with dense mass of small and large rounded yolk globules which are eosinophilic. The zona pellucida and follicular layer consisting of thermal layer is distinctive.

viii. Tertiary Yolk Globule Stage: These stage oocytes are abundantly seen during prespawning and spawning phases and represent the last stage of yolk accumulation. The ooplasm appeared to contain dense mass of yolk globules which are round to oval in shape and strongly acidophilic. The germinal vesicle is conspicuous with homogenous nucleoplasm and peripheral nucleoli.

ix. Migratory Nucleus: The oocytes at this stage have an average diameter compratively increased to 187.77 µm. The follicular layer is distinct with theca, granulosa and zona pellucida layers. The oocyte is fully occupied with dense mass of yolky material.

In addition to these nine oogenetic stages of oocytes there are post ovulatory follicles and few preovulatory atretic follicles in the ovary.

The Seasonal Ovarian Cycle

The ovary *N. notopterus* exhibit seasonal growth and depletion during one year period. The comparison made between gonadosomatic index and the histology of the ovary shows that the ovaries were growing during preparatory and prespawning phases and subsequently spawned during spawning phase and remained inactive during November and December (post spawning phase). Thus the data on the morphological and histological changes in the ovary indicates that the fish *N. notopterus* passed through different phases of the breeding cycle described below:

i. Preparatory Phase: This phase of the breeding cycle in female is characterised by the presence of large number of growing oocytes belonging to the early and late perinucleolus stages. The GSI (OSI) values low during January increased in the subsequent months. The diameter of the oocytes also increases as the oocytes starts accumulation of cortical alveoli.

ii. Prespawning Phase: This phase of the breeding cycle in *N. notopterus* falls during the months of April, May, and June and extend up to July. The GSI (OSI) also increases attaining peak value during July. The histology of the ovary showed the presence of all the stages of oocytes with a large number of oocytes belonging to vitellogenic group. The transformation of oocytes from primary yolk globule stage to secondary yolk globule stage and to tertiary yolk globule stage might be during the month of April - June. The presence of very few migratory nucleous stage oocytes are seen while their number increases during August indicating final maturation (Germinal vesicle breakdown) occurs at the time of spawning.

iii. Spawning Phase: In this phase the GSI (OSI) reduces compared to earlier phase. This decline continuously in subsequent months. The histology of the ovary in the early phase shows the presence of oocytes belonging to earlier stages (late perinucleolus) and oocytes belonging to advanced stages of oocytes. This phase ovary can be classified as matured ovary.

iv. Post spawning Phase: During the months of November and December the ovary in a histological observation consists of the stages belonging to oogonium, chromatin nucleus and early perinucleolus hence, this phase can be catagorised as immature stage

Table showing gross characteristics of gonads during different phases of the reproductive cycle in the fish, *Notopterus notopterus* (Pallas).

Reproductive stage	*Ovaries*	*Testes*
I. Preparatory phase (January-February)	Pale yellowish in colour yolky oocyte are transparent and visible occupying space in the visceral cavity.	Whitish, translucent uneven in size broader at one end.
II. Pre-spawning phase (April-May)	Yellow in colour occupying more space in the abodominal cavity, vitellogenic oocytes clearly visible	Increases in size appear turgid, opaque and pink in colour vascularity increases
III. Spawning phase (August-September)	Attains deep yellow colour, occupies major portion of abdominal cavity ripe ova deep yellowish in colour, yolk leden under abdominal pressure ooses the ripe oocytes	Extensive enlargement in size – bigger with greying pink, under abdominal pressure milt ooses.
IV. Post spawning phase (November December)	Shrunken in size small, translucent few unspent oocytes	Small in size whitish in colour translucent, a symetrical and uneven.

Ovarian Steroid Synthesising Cellular Sites

They can be identified in histological preparation during prespawning and spawning phases.The follicular wall of *Notoptrus notopterus* made up of zona pellucid, follicular epithelium consisting of granulose and thecal cells of growing oocyte. The follicular wall becomes clearly visible only in perinucleolus and cortical alveoli stages of oocytes and gets differentiated as zona pellucid, follicular epithelium and basement membrane in tertiary yolk and migratory nucleus stages. the follicular epithelium could be well differentiated in the secondary yolk stage onwards. Initially few follicular cells are present in young oocytes. When the oocytes grow the follicular epithelium becomes more developed and differentiated. The Histochemical studies have demonstrated the presenclipid drplets in the follicular epithelium. The follicular epithelium of fully vitellogenic oocyte exhibits increased height with sudanophilic reaction. The follicular epitheliumof oocytes undergoing vitellogenesis exhibit an intense sudanophilia in both the cytoplasm and follicular epithelium, thus suggesting the presense of sudanophilic lipid in the follicular epithelium may be related to the process of vitellogenesis and steroidogenesis. The follicular epithelium in the fish, *Notpterus notopterus* gives positive reaction to ascorbic acid,glycogen, alkaline phosphatase and acid phosphatase reactions in the vitellogenic oocytes.

It has been reported in recent years that the developing follicular wall of developing oocytes in different fish species has been shown to play significant role in the growth and differentiation of fish eggs. Various Histochemical studies have suggested that follicular epithelium is involved in the synthesis of protein and lipid during the growth of the oocytes. The other histochemical studies such as enzymes demonstrated in the follicular wall are related to either permeability or steroid biosynthesis.

6. Ovarian Steroid Hormones

a) Estradiol-17β (E-17β): The plasma level of estradiol 17β concentration (pg/ml) values for all four reproductive phases has been determined. Plasma estradiol level gradually increased from post spawning to preparatory phase and marginally declined during prespawning and spawning phases in females whereas in males peak estradiol level is during prespawning and thereafter lower level during postspawning. The higher level of estradiol in females during preparatory phase may be because of preparation of female gonad for vitellogenic activity and in male for spermatogenesis.

b) Progesterone: The male and female *N. notopterus* plasma progesterone concentration (mg/ml) was determined for the entire four reproductive phases. Female progesterone levels were low during post spawning where after it increased significantly during spawning phase. The plasma concentration of progesterone were a higher in females than males. In male the level of the hormone though less in all the phases significant increase was noticed during postspawning phase. In females the progesterone level increased during prespawning and spawning phase may be because, some developed oocytes have been spawned during prespawning and post ovulating follicles increase resulting in higher production of the hormone.

Correlative Tissue Biochemical and Plasma Hormonal Changes

The biochemical changes in tissues such as liver and muscle during different stages of reproductive cycle is correlated with hormonal changes seen during these phases. The pituitary hormones such as FSH, LH and prolactin gradually increases from preparatory phase upto spawning phase and similar response has been noticed for gonadal hormones such as estradiol-17β, testosterone and progesterone in both male and female fish indicating pituitary hormones stimulate the synthesis of gonadal hormones.The hepatic and muscle biochemical content of protein increases during preparatory and spawning phase, the plasma estradiol-17β level also increase during these phases indicating hormonal stimulation for increased protein synthesis needed for vitellogenesis and spermatogenesis. The glycogen level decreases during prespawning and spawning phases in both liver and muscle with an increased level of gonadal hormones indicating utilization of glycogen for reproductive activities (gonadal growth and spawning). The hepatic and muscle lipid level increases during preparatory phase with an increased level of estradiol 17β and testosterone indicating lipid synthesis in both the tissues. The hepatic and muscle cholesterol reduces in the preparatory phase while increases during spawning phase in both the tissues. The progesterone and estradiol and testosterone suggests the hormone stimulation for their synthesis needed for preparation for gonadal growth particularly for vitellogenesis and spermatogenic activities.

The plasma levels of various hormones produced from pituitary and other endocrine glands in relation to seasonal gonadal development and maturation have been reported in several species of fish which has proved to be valuable tool in determining the role of hormones in growth of gonads and involvement of associated tissues such as liver and muscle for reproduction in the fish. These studies also provide information in understanding of endocrine control of reproduction in teleosts. The changes in plasma pituitary hormones levels in fish in relation to gonadal activity have been documented in variety of fishes. Idler and Ng, (1983) have reviewed on teleost gonadotropins stating that efforts to equate mammalian with teleost gonadotropins have not been too successful, some teleost gonadotropins have been found to resemble both mammalian FSH and LH in their amino acid compositions and mammalian FSH and LH in chromatographic behaviour. Certain teleost gonadotropins possessed leutinising hormone like biological activities but in other separation of follicle stimulating hormone like and leutinising hormone like biological activities were not clear cut. The FSH and LH were referred to as vitellogenic hormone and maturational hormone respectively in teleost fishes.

The pituitary hormone (maturational) content was studied in gold fish at different stages of sexual maturation in the annual reproductive cycle, Further, the variation of plasma titers of the gonadotropin was studied with respect to changes in both reproductive status of the fish and the ambient temperature. The results show that the pituitary content of maturational hormone increased with the progress of sexual maturation, the plasma level was contingent on both the sexual status and the ambient temperature with gonadal maturation and high temperature favouring a high plasma level. In the fish *N. notopterus* in the present study plasma FSH and LH have been estimated by applying RIA technique using coat A method. The hormones have been quantitatively detected during four reproductive phases. The hormone FSH may be similar to the vitellogenic hormone and LH may be maturational hormone as reported in other fishes. Since these hormones exhibited seasonal rhythms of changes showing gradual increase of FSH from preparatory phase upto spawning phase the time in which active vitellogenesis and spermatogenesis take place.

While the hormone LH increase during prespawning and remains more or less same during spawning and post spawning phase in male and higher level of the hormone during spawning phase in female fishes indicates their role in activation of steroidogenic cells and spermiation process in males and steroidogenic and spawning process in females.

Gonadotrophins stimulate the 17β-estradiol production of the ovary of *Sarotherodon aureus* and androgen production of the testes of *Cyprinus carpio, Gilliethys mirabilus* and *Pleuroneetes platessa*. The SG-G100 induced a free cholesterol depletion from *Channa punctatus* ovarian tissues incubated *in vitro*. In the fish *N. notopterus* in the present study FSH and LH increases during prespawning and spawning phase, the plasma level of the estradiol and testosterone also increases during these phases indicative of stimulation of gonads for their steroid production. The biochemical changes observed in the liver and muscle during four phase of reproductive cycle may not be the direct effect of these hormones however, they may be acting through the steroid hormones produced from the respective gonads in male and female fish.

Fostier et al., (1983) have reviewed the gonadal steroids in teleosts. A number of workers have reported biosynthesis of steroids from the gonads of variety of fishes and generally steroid hormones such as 17β-estradiol, testosterone and progesterone estimated in the present study in the fish *N. notopterus* have been also identified in number of teleosts during different phases of reproductive cycle. The ovarian steroid hormones in females includes 17β-estradiol, testosterone and progesterone along with other androgens and estrogens. Testicular steroid hormones in males includes testosterone, estradiol 17β and progesterone along with other androgen and estrogens. The steroid hormones such as estradiol 17β, testosterone, and progesterone have been also detected and quantitatively determined by RIA using coat A method in both male and female fish *N. notopterus* in the present investigation during four reproductive phases.

Correlation between seasonal changes in plasma levels of gonadal steroids and gonad condition have been well documented in a number of freshwater fish species. Seasonal variation in serum gonadal steroid and associated changes in the biochemical content of the tissues in the present study shows that there existence of correlation in relation to reproductive cycle in the freshwater fish *N. notopterus*. Steroid hormone levels particularly estradiol-17β and testosterone levels were low in postspawning whle they gradually increased during preparatory and prespawning phases in both the sexes. Seasonal variation of steroid hormone level in an intertidal resting fish, plain fish midshipman (a deep water teleost fish) the plasma levels of testosterone and estradiol-17β were determined and found very low through out the year and peaked during ovaries underwent seasonal recrudescence. Seasonal changes in serum concentration of the testosterone in males and 17β-estradiol in females and triglycerides and cholesterol in both sexes of Copoeta copoeta were determined and it has been observed that seasonal changes in both serum lipids and steroid hormones were associated with reproductive activities. Estrogen synthesis, mainly 17β-estradiol and / or estrone has been found in most teleosts examined. The estradiol-17β is secreted from the gonads in both male and female fish. In general estradiol is responsible for stimulating vitellogenesis and is also secreted by female gonads. The plasma level of this hormone in *N. notopterus* increases during preparatory and prespawning phases reflects the importance of this hormone.

The higher level of this hormone during preparatory phase in female and prespawning phase in male indicates in female its involvement in vitellogenesis and there is also increase in protein content of liver and muscle promoting protein synthesis by this hormone. Estradiol has been reported to stimulate vitellogenesis in teleosts. There is decrease in the level of estradiol during spawning phase compared to preparatory and prespawning phase in *N. notopterus* may possibly indicate that a rapid utilization of the hormone is stimulating vitellogenesis or completion of vitellogenesis as the fish undergoes spawning activity resulting in post ovulatory follicle formation which secrets progesterone. In the fish *Oreochrosmis mossambicus* it was reported that during resting season there are generally low levels of steroid hormones and during ovarian recrudescence, there is an increase in trophic and steroid hormones and suggests that this relates to vitellogenesis. In the fish *N. notopterus* in the present investigation the pituitary hormones FSH, increase with an subsequent increase in the estradiol production during breeding phase (which includes preparatory, prespawning) reflects its involvement in vitellogenesis. Such observation has been also made in the Indian catfish *H. fossilis* showing that such estradiol peak corresponds to a rapid vitellogenic growth phase in the oocytes and the level of estradiol maintaining upto spawning phase for protection and to prevent the oocytes from becoming atretic. There is a good correlation between circulating estradiol 17β and protein content of the tissue (liver and muscle) in the present study that the plasma estradiol-17β in *N. notopterus* paralleled increases of the protein content of tissues, there by confirming a role of estradiol-17β in protein synthesis for vitellogenesis. In *N. notopterus* plasma levels of progesterone in female is higher during spawning phase and lower during post spawning and preparatory phases. This reflects formation of post ovulatory follicles and increase in progesterone level during spawning and also some increase during prespawning phase indicates that some developed oocytes have spawned.

An inverse relationship was found to be seen between plasma level of progesterone and estradiol-17β levels in the female fish *N. notopterus*, when plasma progesterone levels are high, estradiol levels are low. This observation indicates the development of ovarian follicles and recrudescence. Such observation has been noticed in the fish *O. mossambicus*. In the male fish earlier studies have shown that increasing plasma levels of progestin with sperm production and androgen synthesis and stimulation of mitosis of germ cells concerning estrogens, biosynthesis in vitro was also reported in the testis of ambisexual species in the male fish *N. notopterus* plasma levels of progesterone was very low compared to estradiol. However, progesterone increases during post spawning phase and preparatory phase may indicate its involvement in the testicular activities associated with probably androgen synthesis and stimulation of germ cells.

The plasma testosterone levels in *N. notopterus* found to be increased gradually from preparatory phase to spawning phase with decline during post spawning phase indicates its role in gonadal growth particularly for spermatogenesis in male and vitellogenesis in female. The testosterone levels are higher in males than females corresponds to its requirement for spermatogenesis in males than females. The general increase in testosterone from preparatory phase to spawning phase indicate the commencement of spermatogenesis within the testis, such increase level of testosterone in both male and female fish has been reported in the fish, *O. mossambicus* and suggested its need for spermatogenesis in males and growth of gonads in females.

Earlier studies also show that the androgen production from the ovaries has been reported in fish species in vitro studies. In the grey mullet, *Mugil cephalus*, the ovarian production of the ketotestosterone increases with the development of vitellogenesis and then decreases after spawning.

Effects of Steroid Hormones on Tissue Biochemical Contents

The use of steroid hormones as dietary supplements for growth promotion is an accepted practice. A variety of hormones, both natural and synthetic have been evaluated for their anabolic properties and in fish androgen have given more consistent results than estrogens studies pertaining to the administration of various hormones on the biochemical contents of tissues are although available in Indian fishes are not towards the determination of nutritional status of the fishes. Hence, in the present study an attempt is made to understand effects of some important metabolic hormones such as cortisol, testosterone and thyroxin on tissue biochemical changes specially to understand their metabolic role in enhancing the tissue biochemical contents for nutritional purpose. The hepatic and muscle protein is found to be increased in the testosterone treated fish. In cortisol treated fish hepatic protein decreased whereas the muscle protein shows fluctuation. The hepatic glycogen content is decreased in testosterone treated fish whereas it is found to be increased in spawning and post spawning phase. The steroid hormones progesterone and estrogen were injected in two different doses to the fish, *Notopterus notopterus* to observe changes in the GSI and HSI along with testicular and ovarian histological and biochemical changes during pre-spawning phase. The GSI increased in progesterone and combination of estrogen and progesterone treated fish. After progesterone treatment the testicular lumen was becoming empty and estrogen treatment caused accumulation of spermatozoa in the lobular lumen whereas the combination of these steroids caused division of spermatogonial and spermatocytes. The ovarian changes observed after the treatment of these steroids activated vitellogenesis, transforming into vitellogenic oocytes as the hepatic cells became hyper active. The changes in the biochemical contents of protein, cholesterol, glycogen in the liver, ovary and testis has been noticed in response hormonal treatment as there is increase in the ovarian protein,increase in the hepatic, ovarian, testicular cholesterol along with increase in the hepatic glycogen indicating positive relationship between liver and gonads. The results obtained after these steroid hormonal treatment in *Notopterus notopterus* has indicated that liver, ovary and testis are responsive to progesterone and estrogen with their combination because of availability of receptors during pre-spawning period. The progesterone is probably involved in spermiation process and estrogen promotes spermatogenesis in the testis. The progesterone causes oocyte maturation and estrogen increases vitellogenic activity. Various other morphological and biochemical studies shown that the females specific protein – vitellogenin is synthesized and secreted by the liver under the stimulation of estrogen either of ovarian origin during breeding season or exogeneously administered. Induction of final maturation, ovulation and spermiation has been observed by administration of progesterone than testosterone in the fish, Northern Pike. It has been also demonstrated that in male rainbow trout with the initiation of spermiation as there is a drop in the androgen concentration and increase in the plasma concentration of 17 β-hydroxyprogesterone and 17 β-hydroxy 20β-dehydro progesterone. The progesterone levels remained high throught spermiation and positively co-related with sperm volume.

Section of testis showing dividing spermatogonial cells during preparatory phase and sudanophilic Interstetial Leydig cells of the fish, *Notopterus notopterus*.

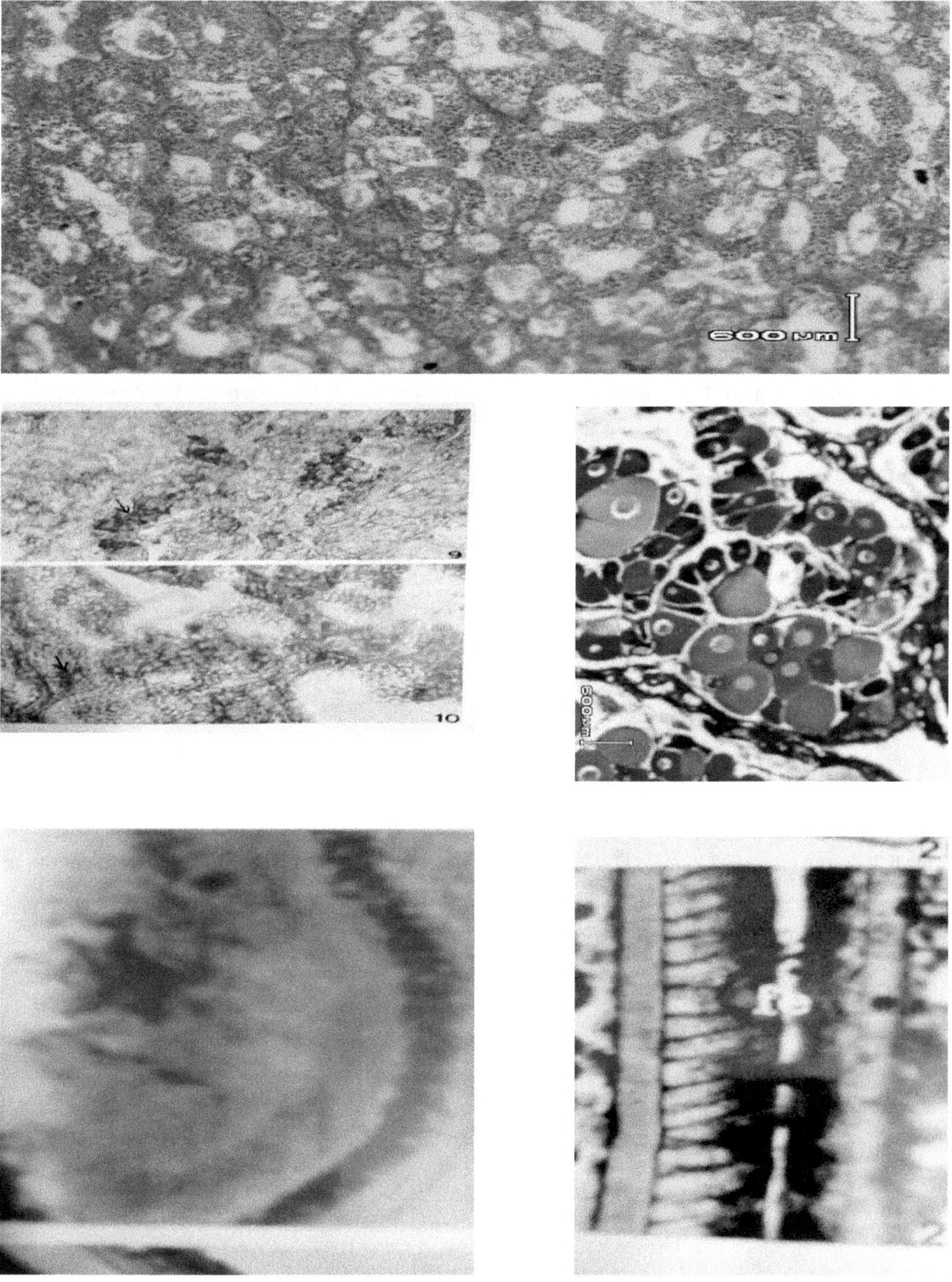

Fig.6.1: Section of ovary showing different stages of oocytes during preparatory phase and sudanophilic follicular cells of the fish, *Notopterus notopterus*

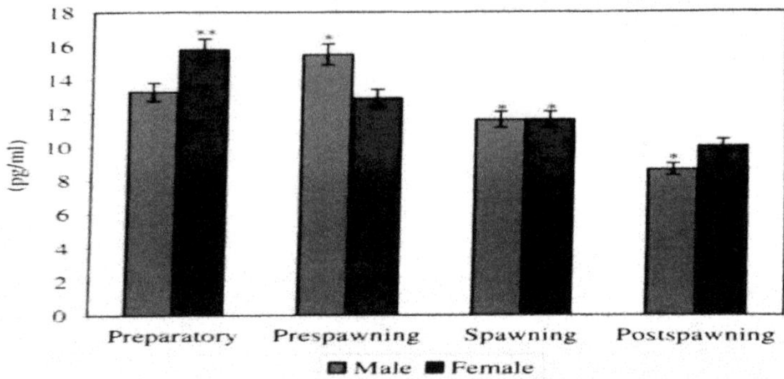

Fig.6.2: Serum estradiol 17β in male and female fresh water fish *N. notopterus* during different phases of the reproductive cycle (pg/ml)

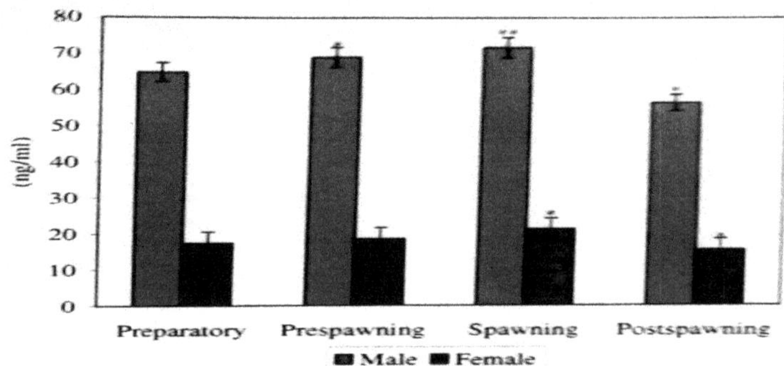

Fig.6.3: Serum testorone in male and female fresh water fish *N. notopterus* during different phases of the reproductive cycle (pg/ml)

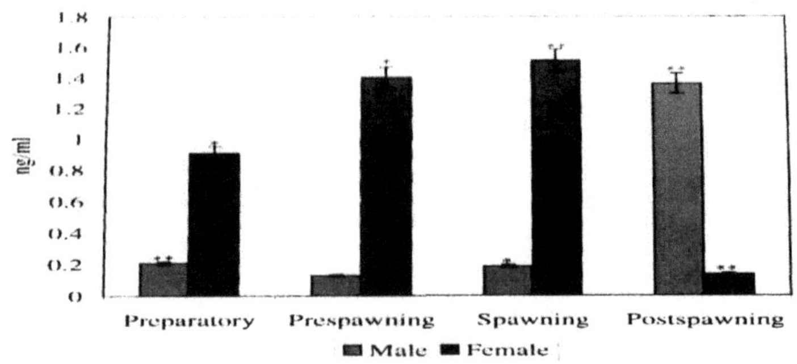

Fig.6.4: Serum progesterone in the female fresh water fish *N. notopterus* during different reproductive phases

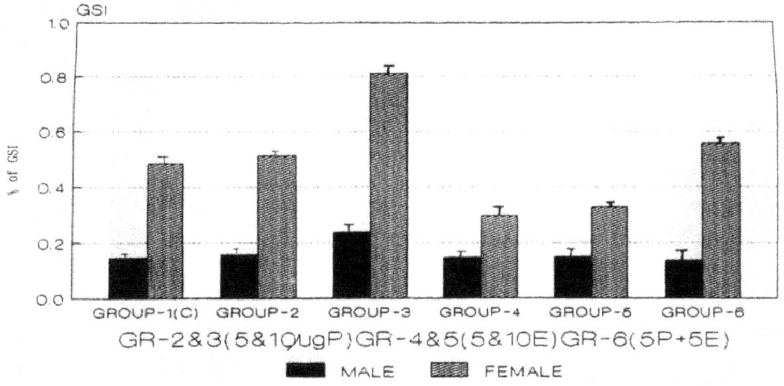

Fig.6.5: Gonadosomatic index of *N. notopterus* in response to hormones

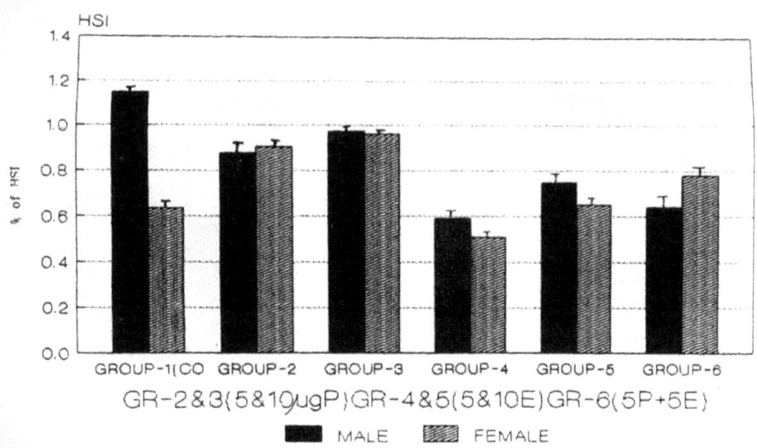

Fig.6.6: Hepatosomatic index of *N. notopterus* in response to hormones

References

Baker, B.I., Keshavanath, P. and Sundararaj, B.I. (1974). The cell types in the adenohypophysis of Indian Catfish *Heteropneustes fossilis* (Bloch). *Cell Tiss. Res.,* 148: 347-358.

Ball, J.N. (1981). Hupothalamic control of the pars distalis in fishes. Amphibians and reptiles. *Gen. Comp. Endocrinol.,* 44: 135-170.

Ball, J.N. and Baker, B.I (1969). The pituitary gland; anatomy and physiology, In: fish physiology (Eds. Hoar, W.S. and D.J. Randall), Vol. II, Academic Press, New York, 1-110.

Bhaskaran, G. and Sathyanesan, A.G. (1982). Tetraped like hypothalamo-hypophysial portal system in the teleost Magalops cyprinoids, (Broussonet), *Gen. Comp. Endocrinol.,* 86:211-219.

Dodd, J.M. and Ker, T. (1963). Comparative morphology and histology of hypothalamo neurohypophysial system. *Symp. Zool. Soc.,* (London), 9:5-27.

Follenius, E., Doerr-Schott, J. and Dubois, M.P. (1978). Immunocytology of pituitary cells from teleoost fishes, *Int. Rev. Cytol.,* 54:193-223.

Gabe, M. (1966). Neurosecretion, Pergamon Press, Oxford.

Gopal Dutt, N.H. (1989). Cyclic changes in the pituitary gland of vertebrates in relation to reproduction. In; Reproductive Cycles of Indian Vertebrates (Ed. S.K. Saidapur) 1-57. Allied Publishers. New Delhi.

Haider, S and Singh, I.J. (1979). An *in situ* study of the hypothalamo-neurohypophysial complex of the freshwater teleost, *Ompak bumaculatus* (Bloch) with a note on the tetrapodan features of its vascularization, *J. Neural. Transmis.,* 45:239-251.

Haider, S. and Sathyanesan, A.G. (1971). Hypothalamo-neurohypophysial complex of the teleost. *Macrognathus aculeatus* (BL). *Acta. Zool.,* 52:235-239.

Haider, S. and Sathyanesan, A.G. (1972a). Comparative study of the hypothalamo neurohypophysial complex in some Indian freshwater teleosts, *Acta. Anat.,* 81:202-224.

Haider, S. and Sathyanesan, A.G. (1972c). An *in situ* study of the hypothalamic neurosecretory system of the freshwater cat fish, *Ailia coila* (Ham). *Acta. Zoologica.,* 53:23-30.

Holmes, R.L. and Ball, J.N. (1974). The pituitary gland. A comparative account, Cambridge University Press, England.

Jorgensen, C.B. and Larsen, L.O. (1967). Neuroendocrine mechanism in lower vertebrates. In: Neuroendocrinology. (Eds. L. Martini and W.F. Ganang), Academic Press, New York. 3:485-521.

Jose, T.M. and Sathyanesan, A.G. (1977). Hypothalamo-hypophysial neurosecretory and vascular system of the teleost, *Labeo rohita* (Ham), *Anat. Anz.,* 142:424-434.

Jose, T.M. and Sathyanesan, A.G. (1977). Pituitary cytology of the Indian carp, *Labeo rohita* (Ham), *Anat Anz.,* 142:410-423.

Joy, K.P. and Sathyanesan, A.G. (1978). Micromorphology and Pituitary cytology of the teleost fish, *Tilapia mossambica* (Peters). Z. Mikrosk. *Anal. Forsch. Leipzig.,* 94:337-344.

Joy, K.P. and Sathyanesan, A.G. (1978). Micromorphology and angioarchitecture of the hypothalamo-neurohypophysial neurosecretory complex of the teleost, *Cirrihinus mrigala* (Ham). *Arch Ital. Anat Embriol.,* Vol. 83, Fasc 3.

Matan Golan, Agnes O.M., Patrice M. and Berta LSivan , Anatomical and functional gonadotrope networks in the teleost pituitary. Sientific reports, 6, 23777; dol: 10.1038/strep 23777 (2016).

Naito K, Suzuki P, Swanson M, Nozaki H, Kawauchi Y Nakai, (1993) Ultrastructural characteristics of two gonadotrophs (GTH I and GTH II cells) in the Pituitary of the Rainbow trout (*Oncorhynchus Mikiss*). Fish hysiol Bichem 11: 291-246.

Peter, R.E., Trudeau, V.L. and Sloley, B.D. (1991). Brain regulation of reproduction in teleosts. Monograph, *Bull-Inst. Zool., Acad Sinica,* 16: 89-118.

Ramaswami, L.S. (1980). Vertebrate neurosecretion – A review. Proc. *Indian Natn. Sci.,* B- 46, Supplement, 1:1-78.

Rodriguez, E.M. (1976). Cerebrospinal fluid as a pathway in neuroendocrine regulation. *J. Endocrinol.,* 71:407-443.

Sahai, S. (1986). Localization of cell types in the adenohypophysis of the pituitary gland of *Puntius ticto* (Ham. Buch), telecast. *Intl. J. Acad. Ichithyol.,* 7(2):11-15.

Sathyanesan, A.G. (1970a). Hypothalamo-hypophysial vascularization and its relationship with the neurosecretory system in the catfish. *Calrias batrachus (L). Zool Beitrage.,* 17:1-11.

Sathyanesan, A.G. (1970b). The pre-opticus nucleus of the freshwater catfist, hypophysial system and vascular system in the teleost, *Rita rita* (Ham). *J. Neural*

Sathyanesan, A.G. (1970c). The nucleus lateralis tuberis of *Clarias batrachus (L). Zool. Beitrage.,* 16:257-262.

Sathyanesan, A.G. and Das R.C. (1975). Relationship between the pre-optico hypophysial system and vascular system the teleost, *Rita rita* (Ham). *J. Neural. Transmis.,* 37:33-42.

Sathyanesan, A.G. and Das, R.C. (1978). Hypothalamo-hypophysial vascularization of teleost, *Puntius sophore* (Ham) with special reference to its tetrapodan features. *Anat. Anz.,* 143:110-119.

Sathyanesan, A.G. and Haider, S. (1971). Tetrapod like features of the hypothalamo-hypophysial vascularization in the air-reathing teleost. *Heteropneustes fossilis* (BL). *Gen. Comp. Enclorinol.,* 17:360-370.

Sathyanesan, A.G. and Jose, T.M. (1973). Structure of the pre-optic neurohypophysial complex of the teleost, *Nangra punctata* (Day), *Anat. Anz.,* 134:230-238.

Sathyanesan, A.G. and Jose, T.M. (1975a). Structure of the neurohypophysis and hypothalamo-hypophysial vascularization in the teleost, *Channa punctatus* (Bloch). Z. Miukrosk. *Ant. Forsch-Leipzig.,* 89(5):824-839.

Sathyanesan, A.G. and Jose, T.M. (1975b). Hypothalamo-hupophysial vascular and neurosecretory links in the teleost, *Bogarius bagarius* (Ham). *Acta. Anat.,* 93:387-389.

Sathyanesan, A.G. and Kulkarni, R.S. (1983). Hypothalamo-hypophysial neurosecretory system and its vasculature in the freshwater catfish. *Mystus vittatus.* Z. Mikrosk. Anat. Forsch. Leipzig., 97(6):1009-1021.

Schreibman, M.P. (1986). The pituitary gland. In; Vertebrate endocrinology, (Eds. Peter, K.T. Pang and Martin, P. Schreibman), 1:11-55.

Scott, D.E., Krobisch-Dudley, G., Pauli, W.K. and Kozlowski, G.P. (1977). The ventricular system in the neuroendocrine systems, III sopra ependymal neuronal networks in the primate. *Cell Tiss. Res.,* 179: 235-254.

Singh, I.J and Haider, S. (1977). Hypothalamo-hypophysial neurosecretory system of the teleost, *Trichogaser fasciatus* (Bloch and Schenider) with a note on its vascularization. Z. Mikrosk. *Anat. Forsch. Leipzig.,* 91(3):509-520.

Sundararaj, V.I. and Vishwanathan, N. (1971). Hypothalamo-hypophysial neurosecretory and vascular system in the catfish, *Heteropneustes fossilis* (Bloch). *J. Comp. Neurol.,* 141:95-106.

Van Oordt, P.G.W.J. and Peute, J. (1983). The cellular origin of pituitary gonadotropins in teleosts. In; Fish physiology, (Eds. W.S. Hoar, D.J. Randall and E.M. Donaldson), Academic Press, New York.9:137-186.

Pituitary Hormones

Bonna-Gallo, A. and Licht, P. (1981). Gonadotropin specificity of *in vitro* testosterone secretion by fish testes. *Gen. Comp. Endocrinol.,* 43, 467-478.

Cambray, J.A. and Bruton, M.N. (1984). The reproductive strategy of a barb, *Barbus anoplus* (Pisces), Colonizing a man made lake in South Africa, *J. Zool. Lond.,* 204: 143-168.

Campbell, C.M. and Idler, D.R. (1976). Hormonal control of vitellogenesis by hypophysectomized winter flounder (*Pseudopleuronectes americanus* Walbaum). *Gen. Comp. Endocrinol.*, 28, 143-150.

Chan, S.T.H. and Philips, J.G. (1969). The biosynthesis of steroids by the gonads of the rice-field cel *Monopterus albus* at various phases during natural sex reversal. *Gen. Comp. Endocrinol.*, 12, 619-636.

Colombo, L., Colombo Belvedere, P., Bieniarz, K., Epler, P. (1982). Steroid hormone biosynthesis by the ovary of carp, *Cyprinus carpio* L. during wintering. *Comp. Biochem. Physiol.*, 72B, 367-375.

Cook, A.F. and Peter, R.E. (1980). Plasma clearance of gonadotropin in goldfish, *Carassium auratus,* during the annual reproductive cycle. *Gen. Comp. Endocrinol.*, 42, 76-90.

Crim, L.W. and Idler, D.R. (1978). Plasma gonadotropin, estradiol and vitellogenin and gonad phosvitin levels in relation to the seasonal reproductive cycles of female brown trout. *Ann. Biol. Anim. Biochem. Biophys.*, 18, 1001-1005.

De Valming V.L., Wiley, H.S., Delahunty, G. and Wallace, R.A. (1980). Gold fish (*Carassius auratus*) vitellogenin: Induction, isolation, properties and relationship to yolk proteins. *Comp. Biochem. Physiol.*, 67B, 613-623.

Duggan, R.T. and Bolton, J.P. (1982). Androgen production by the plaice, *Pleuronects platessa,* Effects of gonadotropic preparations and metyrapone. *Gen. Comp. Endocrinol.*, 46, 392 (Abstr. 135).

Fontaine, Y.A. (1975). Hormones in Fishes. In: "Biochemical and Biophysical perspective in marine biology" (Eds. D.C. Malins and J.R. Sargent) Vol. 2, 139-212, Academic Press, London and New York.

Fostier, A., Jalbert, B., Billard, R., Breton, B. and Zohar, Y. (1983). Gonadal steroids In: "Fish physiology". Vol. IXA (Eds. W.S. Hoar, D.J. Randal and E.M. Donoldson), pp. 277-372, Academic Press, New York, London.

Gopal Dutt, N.H. (1989). Cyclic changes in the pituitary gland of vertebrates in relation to reproduction. In: Reproductive cycles of India vertebrates (Ed. Saidapur S.K.) Allied publishers, New Delhi, pp. 1-57.

Huang, F.L. and Chang, Y.S. (1980). The gonadotropic stimulation of androgen production on carp testis *in vitro. Proc. Natl. Sci. Counc., Repob.* China, 4, 392-400.

Idler and Ng (1983). Teleost gonadotropins: Isolation, Biochemistry and Function. In: "Fish Physiology", Vol. IXA (Eds. W.S. Hoar, D.J. Randal and E.M. Donoldson), pp. 137-221 Academic Press, New York.

Kime, D.E. (1979b). The effect of temperature on the testicular steroidogenic enzymes of the rainbow trout, *Salmo gairdneri. Gen. Comp. Endocrinol.*, 39, 290-296.

Lambert, J.G.D. (1978). Steroidoenesis in the ovary of *Brachydanio rerio* (Teleostei). In: "comparative endocrinology" (P>J. Gillard and H.E. Boer, Eds.), pp. 65-68, Elsevier / North Holland Biomedical Press, Amsterdam.

Mukherjee, D. and Bhattacharya, S. (1981). A sensitive and easy bioassay for teleost gonadotropin depending on the ovarian free cholesterol depletion *in vitro. Gen. Comp. Endocrinol.,* 45, 249-255.

Peter, R.E. (1981). Gonadotropin secretion during reproduction cycles in teleosts: influences of environmental factors, *Gen. Comp. Endocrinol.,* 45, 294-305.

Scott, A.P., Bye, V.J. and Baynes, S.M. (1980). Seasonal variations in sex steroids of female rainbow trout *Salmo gairdneri* Richardson. *J. Fish. Biol.,* 17, 587-592.

Sisneros, J.A., Forlano, P.M., Knapp, R., Bass, A.H. (2004). Seasonal variation of steroid hormone levels in an intertidal-nesting fish, the vocal plainfin midshipman. *Gen. Comp. Endocrinol.,* 136(1): 101-116.

Smith, C.J. and Haley, S.R. (1988). Steroid profiles of the female tilapia *Oreochomis mossambicus,* and correlation with oocyte growth and mouth brooding behaviour. *Gen. Comp. Endocrinol.,* 69(1), 88-98.

Sundararaj, B.L. and Goswami, S.V. (1968). Effects of estrogen, progesterone and testosterone on the pituitary and ovary of cat fish, *Heteropneustes fossils. J. Exp. Zool.,* 169, 211-228.

Theofan, G. (1981). The *in vitro* synthesis of final maternal steroids by ovaries of brook trout *Salvelinus fontinalis* and yellow perch *Perca flavescens.* Ph.D. Thesis, University of Notre Dam, Notre Dam, Indiana.

Van Den Hurk, R., Lambert, J.G.D. and Peute, J. (1982). Steroidogenesis in the gonads of rainbow trout fry *Salmo gairdneri* before and after the on set of gonadal sex differentiation. *Reprod. Nutr. Develop.,* 22, 413-426.

Vander Merwe, W., Van Vurenjhj and Vermaak, J.F. (1987). Cyclic changes of steroid production activity in the ovary and estradiol levels in the blood plasma of the mudfish, *Labeo capensis. S. Afr. J. Zool.,* 22(3), 213-217.

Pancreas

Bonner-Weir S. and Weir, G.C. (1979).The organization of endocrine pancreas: A hypothetical unifying view of the phylogenetic differences. *Gen.Comp. Endocrinolology,* 38,28-37.

Dubois,P. (1989). Ontogeny of the endocrine pancreas. *Honn.Res.,*32,53-60.

Epple,A. and Brinn, J.E. (1975). Islets histophysiology,Evolutionary correlations. *Gen Comp.Endocrinology* ,27,320=349.

Epple,A. and Brinn, J.E. (1986).Pancreatic islets.In: Vertebrate Endocrinology: fundamentals and biochemical interpretations Vol.2, "morphological correlations" P.K.T.Pang and M.T.Schreibman (Eds.), Academic Press,Orlandopp, 279-317.

Epple,A. and Brinn, J.E. (1987).The comparative physiology of the pancreatic islets. springer Verlag,Berlin.

Epple,A.,. Brinn, J.E. and Young, J.B. (1980). Evolution of pancreatic islet functions. In : Evolution of vertebrate endocrine systems. P>K>T>Pang and A. Epple (Eds.), Texas Tech.Univ.Press ,Lubbock, pp.269-321.

Gutierrez, J.,Fernandez, J., Carrillo, M.,Zanuy,S. and Planas, J.,(1987). Annual cycle of plasma insulin and glucose of sea bass, *Dicentrarchus labrax*,L. *Fish Physiology and Biochemistry*,4,3,137-141.

Le- Dourian,N (1988). On the origin of pancreatic endocrine cells. 53, 169-171.

Ottolenghi, C.Punvian, A. C., Baruffaldi, A. and Brighenti,L. (1982). *In vivo* effects of insulin on carbohydrate metabolism of cat fish, (Ictalurus). *Comp.Biochem. Physiol.*72 A, 35-41.

Plisteskaya, M.Erika (1990). Endocrine pancreas of teleost fish ; A model for interaction of islet hormones. *Journal of Expt. Zoolgy*, 256 ,54,53=57.

Thorpe, A. and Ince, B.w. (1974). The effects of pancreatic hormones, catecholamines and glucose loading on blood metabolites in the northern pike. *Gen.Comp. Endocrinol.* 23, 29-44.

Pineal Gland

Bhattacharya, S.,Day,R.,Basu, A.,Maitra, S.K. and Banerji,T.K. (2003). The structure of pineal complex in common Indian teleost, Catla catla : Evidence for pineal induced inhibition of testicular function within an annual reproductive cycle. *Endocrinol. Res.*, 29 (2), 141-156.

Bollet,V., Ali, M. A.,Lapointe, F.J.,falcon, J. (1996). Rhythmic melatonin secretion in different teleost species :an in vitro studies. *Journal of Comp.,Physiol.,Part*-B,165, 677-683.

Ekstrom, P. and Meissl, H. (1997). The pineal organ of teleost fishes. *Review in fish biology and fisheries.* 7, 199-284.

Renuka, K.,Joshi, B.N. (2010). Melatonin induced changes in ovarian function in the fresh water fish, *Channa punctatus. Gen. Comp. Endocrinol* 1,165(1),42-46.

Sastry, V.K.S., and Sathyanesan, A.G. (1981). Comparative study of pineal complex of nineteen species of Indian fresh water fishes. *Journal of Hirnfrosch*, 22, 327-340.

Gonads

Billard, R., Fostier, A., Weil, C. and Breton, B. (1982). Endocrine control of spermatogenesis in teleost fish, *Can. J. Fish Aquatic Sci.*, 39: 65-79.

Bohemen G. van and Lambert J.G.D. (1982). Estrogen synthesis in relation to estrone, estradiol and vitellogenin plasma levels during the reproductive cycle of the female rainbow trout, *Salmo gairdneri. Gen Comp Endocrinol.*, 45: 105-114.

Gokhale, S.V. (1957). Seasonal histological changes in the gonads of the whiting (*Gadus merlangust*) and the Norway pout (*G. esmarkii*) Nilsson, *Indian J. Fish.*, 4: 92-112.

Guraya, S.S. (1994) Gonadal development and production of gametes in fish. *Proc. Indian Natn. Sci. Acad.* B60(1), 15-32.

Guraya, S.S. (1999). Comparative testicular biology in animals Oxford and IBH Publishing Co. Pvt. Ltd., New Delhi, Calcutta.

Malhotra, Y.R., Jyothi, M.K. and Gupta, K. (1989). Reproductive cycles of Indian Vertebrates (Saidapur, S.K. Ed.), Allied Publishers, Ltd., New Delhi, pp. 58-105.

Nagahama, Y. (1983). The functional morphology of teleost gonads. In fish physiology Vol. IX: Reproduction Part A (Hoar, W.S., Randall, D.J. and Donaldson, E.M. Eds.) Academic Press, New York, pp. 223-276.

Nagahama, Y. (1986). 'Testis' Ch. 13. In: Vertebrate endocrinology: Fundamentals and biomedical implications, Morphological considerations (Pang, P.K.T. and Schreibman, M.P. Eds.), Academic Press, New York, 1, pp. 399-437.

Nagahama, Y. (1990). Endocrine control of oocyte maturation in teleosts. In: Progress in comparative endocrinology. (Epple, A., Scanes, C.G. and Stetson, M.H. Eds.), Wiley-lis publication, New York, 342: pp. 385-392.

Nayyar, S.K. and Sundararaj, B.I. (1970). Seasonal reproductive activity to the testes and semina vesicle of the cat fish, *Heteropneustes fossilis* (BI), *J. Morph.*, 130: 207-226.

Peter, R.E. and Crim, L.W. (1979). Reproductive endocrinology of fishes: Gonadal cycles and gonadotropin in teleosts. Annu. Rev. Physiol., 41: 323-335.

Singh, S. and Srivastav Ajai, K. (1990). Changes in the serum calcium and phosphate levels in relation to the annual reproductive cycle of the freshwater catfish, Hetero-pneustes fossilis. *Boletim de Physiologie Animale,* 14: 81-86.

Adrenal and Corpuscles of Stannius

Assem,H. and Hanke, W. (1984). A comparision between the effects of cortisol and prolactin on the euryhaline Tilapia (*Sarotherodon mossambicus*). *Zool.Jb. Physiol.*,88 ,423- 431.

Audit,C., FitzxGerald,G. J..and Guderley, H. (1986).Photoperiod effects on plasma cortisol levels in *Gastersteus aculeatus..Gen.Comp. Endocrinol., 61,76- 81.*

Barton, B.A.,Schreck, C.B. and Barton, L.D. (1987). Effects of chronic cortisol administration and daily acute stress on growth, physiological conditions and stress responses in juvenile rainbow trout. *Dis Aquat. Org.* 2, 173- 185.

Boujard, T. Leatherland, J.F. (1992). Circadian rhythm and feeding time in fishes. *Environ. Boil. Fish.*, 35, 109-131.

Brann , D.W. and Mahesh, V.B. (1991). Role of corticosteroids in female reproduction. *FASEB J.* 5, 2691- 2698.

Chan, D.K.O. and Woo, N.Y.S. (1978). Effect of cortisol on the metabolism of the eel, *Anguilla japonica) gen. Comp. Endocrinol. 35, 205-215.*

Chakrabarti, P. (2014). Histological features of interregnal and chromaffin cells in relation to seasonal activities in *Notopterus notopterus* (Pallas 1769). *Iran J. Ichthyol.* 1 (3), 206-213.

Cook A.F.,Stacey, N.E. and Peter , R.E. (1980). Periovulatory changes in serum cortisol levels in the gold fish, *Carassius auratus. Gen. Comp. Endocrinol.* 40, 507-510.

De Jasus ,E.G.T., Hirano, T. and Innui, Y. (1991). Changes in cortisol and thyroid hormone concentrations during early development and metamorphosis in the Japanese flounder, *Parali chthys olivaceus. Gen. Comp. Endocrinol.* 85, 55-61.

Diang, Lim, E.H. and Lam, T.A. (1994). Cortisol induced hepatic vitellogenin mRNA in *Oreochromis aureus* (Steindachner). *Gen.Comp. Endocrinol.* , 96, 276- 287.

Hathway, C.B., and Epple, A. (1989). The sources of plasma catecholamines in American eel, *Anguilla rostrata. Gen. Comp.Endocrinol.* 74, 418-430.

Kulkarni, R.S. and Sathysanesan , A.G. (1978). Cytochemical and histoenzymological studies on the adrenal of the teleost, *Puntius sophore* (ham). Z.Microsk Anat Forsch., 92, 529-540.

Kulkarni, R.S. and Sathyanesan , A.G. (1982). Histochemistry and histoenzymology of the adrenal of the teleost, *Mystus vittatus. Biol.Bull. India.*, 3 (3), 131-139.

Larson, A.L. (1973). Metabolic effects of epinephrine and norepinephrine in the eel, *Anguilla anguilla. Gen. Comp. Endocrinol.* 20, 115.

Mommsen, T.P., Vijayan, M.M. and Moon, T.W. (1999). Cortisol in teleosts : Dynamics, mechanisms of action and metabolic regulation. *Rev. Fish Bio*l. 9, 211-268.

.Nichols, D. J. and Weisbart, M. (1984). Plasma cortisol concentrations in Atlantic salmon, Salmo solar: episodic variations, diurnal change and short term response to adrenocorticotropic hormone. *Gen. Comp. Endocrinol.* , 56, 169-176.

Pickering ,A.D. and Pottinger , T.G. (1983). Seasonal and diel changes in plasma cortisol levels of the brown trout, *Salmo trutta. Gen. Comp. Endocrinol.* 49, 232-239.

Vijayan M.M. and Leatherland, J.F. (1989). Cortisol induced changes in plasma glucose, protein and thyroid hormone levelsand liver glycogen content of coho salmon (*Oncorhynchus kisutch* Walbaum.). *Can. J. Zool.*,67, 2746- 2750.

Corpuscles of Stannius

Belsare, D.K. (1973). Comparative anatomy and histology of the corpuscles of Stannius in teleosts. *Zeitschrift für Mikmskopisch-Anatomische Forschung (Leipzig)* 87 445-456.

Bhattacharyya, T.K., Butler, D.G. and You Son J.H. (1982). Ultra structure of the corpuscles of Stannius in The garpike (*Lepisosteus platyrhynchus*). *Gen. Comp. Endocrinol.*, 46: 29-41.

Bjornsson B. Th., Haux C., Forlin L. and Deftos L.J. (1986). The involvement of calcitonin in the reproductive physiology of the rainbow trout. Journal of Endo. 108: 17-23.

Butler, D.G. and Ordit, G.Y. (1995). CS and blood flow regulation in freshwater North American eels, *Anguilla rostrata,* J. Endocrinol., 145, 181-194.

Carpenter, S.J. and Heyl, H.L. (1974) Fine structure of the corpuscles of Stannius of Atlantic salrnon during freshwater spawning journey. *General and Comparative Endocrinology* 23 212-223.

Cohen, R.S., Pang, P.K.T. and Clark, N.B. (1975). Ultrastructure of the Stannius corpuscles of the killifish, *Fundulus heteroclitus.* and its relation to calcium regulation. *General and Comparative Endocrinology* 27 413-423.

Firoz Ahmad M. (2004). *Current Science,* Vol. 86, No. 2, 25 January.

Fujita, H. and Honma, Y. (1967). On the fine structure of corpuscles of Stannius of the eel, *Anguilla japonica. Z. Zellforsch. Mikrosk. Anat.* 77, 175-187.

Garrett, F.D. (1942). The development and phylogeny of the corpuscles of Stannius in ganoid and teleostean fishes_ *Journal of Morphology* 70 41-67.

Heyl, H.L. (1970). Changes in the corpuscles of Stannius during the spawning journey of Atlantic salmon *(Salmo salar). Gen Comp Endocrinol* 14:43-52.

Hirano, T. (1989). The corpuscles of Stannius. In *Vertebrate Endocrinology: Fundamentals and Biomedical Implications,* vol. 3 (ed. P. K. T. Pang andM. P. Schreibman), pp. 139-170. New York:

Krishnamurthy, V.G. (1976). Cytophysiology of corpuscles of Stannius. *Int. Rev. Cytol.,* 46: 177-249

Krishnamurthy, V.G. and Bern, H.A. (1969). Correlative histologic study of the corpuscles of Stannius and the juxtaglomerular cells of teleost fishes. *Gen. Comp. Endocrinol.,* 13:313-335.

Krishnamurthy, V.G. and Bern, H.A. (1971). Innervation of the corpuscles of Stannius. *General and Comparative Endocrinology* 16 162-165.

Nadkarni, V.B. and Gorbman, A. (1966). Structure of the corpuscles of Stannius in normal and radiothyroidectomized chinook finger lings and spawning pacific salmon. *Acta. Zool. Stockholm,* 47: 61-66.

Subhedar, N. and Rao, P.D.P. (1979). Seasonal changes in the corpuscles of Stannius and the gonads of the catfish, *Heteropneustes fossilis* (Bloch). *Z mikroskanat Forsch* (Leipzig) 93:74-90.

Swarup, K. and Ahmed, N. (1978). Studies of CS in relation to calcium and sodium rich environments, *Natl. Acad. Sci. Lett.,* 1: 239-240.

Swarup, K., Srivastav, S.P. and Srivastav Ajai, K. (1986). Seasonal changes in the structure and behaviour of Stannius corpuscles and Serum Calcium level of *Clarias batrachus* in relation to the reproductive cycle.

Unsicker, K., Polonius, T., Lindmar, R., Loffelholz K. and Wolf, U. (1977). Catecholamines and 5-hydroxytryptamine in corpusus of Stanus of the solmonid *Solmo irideus* L.: A Study correlating electron microscopical, histochemical and Chemical findings. *Gen. Comp. Endo.,* 31: 121-132.

Wendelaar Bonga S.E. Vander Meij, J.C.A. and Pang, P.K.T. (1980). Evidence for two secretary cell types in the Stannius bodies of the teleost *Fundulus heteroclitus* and *Carassius auratus. Cell Tissue Res.,* 212: 295-306.

Wendelaar Bonga, S.E. and Greven, J.A. (1975). A second cell type in Stannius bodies of two euryhaline teleost species. *Cell. TissueRes.* 159, 287-290.

Wendelaar Bonga, S.E. and Pang, P.K.T. (1986). Stannius corpuscles. In *Vertebrate Endocrinology, Fundamentals and Biomedical Implications* (ed. P. K. T. Pang and M. P Schreibman), pp. 439-464. New York: Academic Press.

Wendelaar Bonga, S.E. and Pang, P.K.T. (1991). Control of calcium regulating hormones in the vertebrates: Parathyroid hormone Calcitonin. Prolactin and Stanniocalcin; *Int. Rev. Cytol.,* 128: 138-213.

Wendelaar Bonga, S.E., Greven, J.A.A. and Veenhuis, M. (1976). The relationship between the ionic campaign ion of the environment and the secretary activity of the endocrine cell types of Stannius corpuscles in the teleost *Gasterosteus aculeatus. Cell Tissue Res.,* 175: 297-312.

Wendelaar Bonga, S.E., Greven, J.A.A. and Veenhuis, M. (1977). Vascularization, innervation, and ultrastructure of the endocrine cell types of Stannius corpuscles in the teleost. *Gasterosteus aculeatus. Journal of Morphology* 153 225-244.

www.ingramcontent.com/pod-product-compliance
Ingram Content Group UK Ltd.
Pitfield, Milton Keynes, MK11 3LW, UK
UKHW021956270726
14060UKWH00002B/538

9 789387 057586